Neues verkehrswissenschaftliches Journal

Ausgabe 35

3D-Printed Scale Model for Detection of Railway Wheel Flats using Augmented Vibration Data from Axle Box

Von der Fakultät Bau- und Umweltingenieurwissenschaften

der Universität Stuttgart zur Erlangung der Würde eines

Doktors der Ingenieurwissenschaften (Dr.-Ing.) genehmigte Dissertation

Vorgelegt von

Eui-Youl Kim

aus Daegu, Südkorea

Hauptberichter:	Prof. Dr.-Ing. Ullrich Martin
Mitberichter:	Prof. Dr.-Ing. Prof. E.h. Peter Eberhard
	apl. Prof. Dr.-Ing. habil. Yong Cui
Tag der mündlichen Prüfung:	5. Juli 2024

Institut für Eisenbahn- und Verkehrswesen der Universität Stuttgart

2024

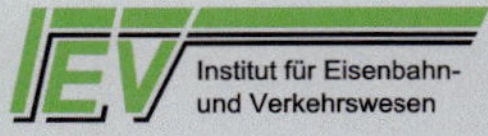

Titelbild: Eui-Youl Kim

Verlag: BoD · Books on Demand GmbH, In de Tarpen 42, 22848 Norderstedt

bod@bod.de

Druck: Libri Plureos GmbH, Friedensallee 273, 22763 Hamburg

Printed in Germany

ISBN 978-3-7597-8819-1

Preface

Wheel flats are typical and relatively common defects on railway vehicles, which are caused in par-ticular by completely or partially blocked wheels, even if the braking system is only malfunctioning for a short time. A small wheel flat can itself impair the effectiveness of the braking system, so that the wheel flat subsequently becomes larger. Wheel flats not only impair the movement of the vehi-cle, but also lead to rail breakage, especially when outside temperatures are low and there is high tensile stress in the rails. These impairments to the infrastructure not only cause high repair costs, but also result in a significant restriction of the use of the infrastructure.

Accordingly, there is great interest in the early and reliable detection of wheel flats and their exact position, even in long trains with a large number of axles. In the past, a large number of people were available to observe the train when a train traveled along the line. These trained personnel were able to detect wheel flats acoustically at an early stage and, by counting the number of vehi-cles remaining until the end of the train, identify the exact position of the wheel flat in the train.

Due to the technical development of the last few decades, there are now relatively few personnel stationed directly at the tracks, and the number of people trained to observe trains has decreased significantly overall. This is where the present dissertation comes in, using AI-supported algorithms to determine wheel flats at an early stage from the vibrations recorded on the axle box. In order to achieve a sufficiently high level of detection, the AI-supported algorithms require suitable training data, which in practice is very difficult and expensive to obtain.

The main research result of this work is a new AI-supported method for the early detection of wheel flats on railway wheels based on the vibrations at the axle box, assessed with AI-supported algorithms whose training data is artificially generated and additionally verified with a test bench model on a scale of 1:10. With the findings of this dissertation and the development of the method for detecting wheel flats on the wheels of railway vehicles, taking into account the limited data availability in practice, a remarkable scientific result was achieved for more efficient maintenance of vehicles and infrastructure while reducing operational restrictions, which also makes another important contribution to the use of AI methods in this area.

Stuttgart, August 2024

Ullrich Martin

Acknowledgements

I would like to express my sincere gratitude to the three professors who have guided me throughout this dissertation.

Prof. Dr.-Ing. Ullrich Martin, thank you for helping me stay on track and encouraging me to look beyond my focus on vehicles to see the broader context of infrastructure and railway operations. I also appreciate how you facilitated opportunities for me to connect with IEV and VWI colleagues and understand their work. Your advice that "sometimes simple is more effective than complex" will always stay with me.

apl. Prof. Dr.-Ing. habil. Yong Cui, I am truly thankful for his guidance and support during the more challenging moments of the projects I worked on. Beyond the research itself, he encouraged me to reflect on what it takes to develop the essential qualities of an independent researcher, giving me the opportunity to explore different approaches and learn through trial and error in the projects.

Prof. Dr.-Ing. Prof. E.h. Peter Eberhard, thank you for the sharp questions and guidance that pushed me to reconsider the fundamental aspects of vehicle design and dynamics, and how to achieve a high similarity between reality and a scale model.

While pursuing my master's degree on another floor of the same building, I had hoped that after graduation, I would be able to continue my doctoral studies somewhere on this campus, ideally within the same building. Fortunately, I was given the opportunity to continue my doctoral studies while working between the offices of IEV, VWI, and CDFEB. I am grateful for the space provided to me in these offices and for the opportunities they offered, allowing me to connect and communicate with IEV and VWI colleagues, which helped me further develop my research and ideas.

I appreciate the direct and indirect feedback and advice from everyone in these offices. Their insights were invaluable to the completion of this dissertation. Additionally, I am thankful for the help I received in daily life from everyone over the past five years. Although I couldn't always express my gratitude properly because of my poor German speaking, I deeply appreciated their support and warm smiles.

Lastly, I owe a special thank you to my parents, Dong-Han Kim and Hae-Ja Kim, and my brother, Joong-Yeol Kim. Their unwavering support, love, and belief in me gave me the strength to overcome challenges and persevere through difficult times.

Eidesstattliche Erklärung

Hiermit erkläre ich, dass ich diese Arbeit selbständig verfasst und keine anderen als die von mir angegebenen Quellen und Hilfsmittel verwendet habe.

Stuttgart, den 05.07.2024   Eui-Youl Kim

Table of Contents

List of Figures

3D-Printed Scale Model for Detection of Railway Wheel Flats using Augmented Vibration Data from Axle Box

List of Tables

3D-Printed Scale Model for Detection of Railway Wheel Flats using Augmented Vibration Data from Axle Box

Abstract

As applications for data-driven defect detection in the railway industry increase, the demand for high-quality data that can be characterised in terms of quantity, variety and velocity is also increasing, as data quality has a significant influence on the results. Relying solely on field experiments to obtain all the necessary data is not often feasible due to the practical constraints of quickly obtaining large amounts of data for various defect conditions. Therefore, there is a significant need for methodologies that reduce reliance on field experiments for wheel flat detection. This study focuses on finding and implementing alternatives to physically generating wheel flat-induced vibration data in the laboratory, and digitally overcoming the time-consuming difficulties of data collection in such an experiment-based approach.

This study aims to develop an efficient methodology that minimizes reliance on field experiments by designing and constructing a scale model of the 'Vehicle on Track' type using Fused Deposition Modelling (FDM) 3D-printing and enhancing vibration data with a Long Short-Term Memory (LSTM)-based generative model in the data acquisition process. The specific objectives include designing a flexible wheelset, modelling 3D wheel flat geometries, improving data quality through a LSTM-based generative model optimised by a genetic algorithm, and classifying these 3D wheel flat geometries using a Convolutional Neural Network (CNN) model optimised by a grid search algorithm.

As a result, the research successfully led to the construction of a 3D-printed scale model that qualitatively implements the mechanism of defect-induced vibration caused by wheel flats and a flexible wheelset in a target railway vehicle and ballastless track. The LSTM-based generative model effectively generated synthetic vibration data with high similarity to real vibration data and contributed to improvements in data volume and velocity. Moreover, the CNN-based classification model, trained on both real and synthetic vibration data, demonstrated enhancements in training stability and accuracy in classifying different 3D-wheel flat geometries.

Key findings are as follows: In the design and construction of a scale model, 3D-printing technology is very helpful. Advanced FDM 3D-printing and materials allow a high degree of freedom in design and construction in a laboratory environment and no durability issues in dynamic testing. The first bending mode of a flexible wheelset dominantly affects vibration data in a realistic way. In studies for wheel flat detection, this allows for a limited similarity link between a real railway vehicle and a scale model. Synthetic vibration data generated by a generative model is effective in improving the classification performance of a classification model. This demonstrates the ability of a generative model to replace some of the experiments and is expected to contribute to time and cost savings. For both generative and classification models, hyperparameter optimisation is critical in their implementation.

In conclusion, the findings of this study are expected to minimise the reliance on field experiments and allow for physically generating and digitally augmenting vibration data quickly and easily in the laboratory and for providing high-quality data for data-driven approaches while saving time and money compared to field experiments. Based on the flexibility, speed, and low cost of FDM 3D-printing, as a practical testing platform, the scale model can be flexibly scaled up for defect detection or other applications for vehicle and infrastructure. This scalable model is expected to greatly accelerate the process of rapidly testing and refining ideas in the preliminary research phase, enhancing research to optimise railway operations and maintenance even within the constraints of field experiments.

Kurzfassung

Da die Anwendungen für datengesteuerte Fehlererkennung in der Eisenbahnindustrie zunehmen, steigt auch die Nachfrage nach qualitativ hochwertigen Daten, die hinsichtlich Menge, Vielfalt und Geschwindigkeit charakterisiert werden können, da die Datenqualität einen signifikanten Einfluss auf die Ergebnisse hat. Sich allein auf Feldexperimente zu verlassen, um alle notwendigen Daten zu erhalten, ist oft nicht machbar aufgrund der praktischen Einschränkungen, schnell große Datenmengen für verschiedene Defektbedingungen zu sammeln. Daher besteht ein bedeutender Bedarf an Methodologien, die die Abhängigkeit von Feldexperimenten für die Erkennung von Flachstellen reduzieren. Diese Studie konzentriert sich darauf, Alternativen zum physischen Generieren von durch Flachstellen induzierten Vibrationsdaten im Labor zu finden und die zeitaufwändigen Schwierigkeiten der Datensammlung in einem solchen experimentbasierten Ansatz digital zu überwinden.

Diese Studie zielt darauf ab, eine effiziente Methodologie zu entwickeln, die die Abhängigkeit von Feldexperimenten minimiert, indem ein Maßstabsmodell des Typs „Vehicle on Track" mittels Fused Deposition Modelling (FDM) 3D-Druck entworfen und gebaut wird und Vibrationsdaten mit einem auf Long Short-Term Memory (LSTM) basierenden generativen Modell im Datenerfassungsprozess verbessert werden. Die spezifischen Ziele umfassen das Design eines flexiblen Radsatzes, die Modellierung von 3D-Flachstellen-Geometrien, die Verbesserung der Datenqualität durch ein mit einem genetischen Algorithmus optimiertes LSTM-basiertes generatives Modell und die Klassifizierung dieser 3D-Flachstellen-Geometrien unter Verwendung eines mit einem Grid-Suchalgorithmus optimierten Convolutional Neural Network (CNN) Modells.

Infolgedessen führte die Forschung erfolgreich zum Bau eines 3D-gedruckten Maßstabsmodells, das qualitativ den Mechanismus von durch Flachstellen und einem flexiblen Radsatz verursachten defektinduzierten Vibrationen in einem Ziel-Eisenbahnfahrzeug und schotterlosem Gleis implementiert. Das auf LSTM basierende generative Modell generierte effektiv synthetische Vibrationsdaten mit hoher Ähnlichkeit zu echten Vibrationsdaten und trug zur Verbesserung des Datenvolumens und der -geschwindigkeit bei. Darüber hinaus zeigte das CNN-basierte Klassifikationsmodell, das mit echten und synthetischen Vibrationsdaten trainiert wurde, Verbesserungen in der

Trainingsstabilität und Genauigkeit bei der Klassifizierung verschiedener 3D-Flachstellen-Geometrien.

Die wichtigsten Erkenntnisse sind wie folgt: Bei der Gestaltung und Konstruktion eines Maßstabsmodells ist die 3D-Drucktechnologie sehr hilfreich. Fortgeschrittener FDM 3D-Druck und Materialien ermöglichen einen hohen Grad an Freiheit in Design und Konstruktion in einem Laborumfeld und keine Haltbarkeitsprobleme bei dynamischen Tests. Der erste Biegemodus eines flexiblen Radsatzes beeinflusst die Vibrationsdaten auf realistische Weise dominierend. In Studien zur Flachstellenerkennung ermöglicht dies eine begrenzte Ähnlichkeitsverbindung zwischen einem echten Eisenbahnfahrzeug und einem Maßstabsmodell. Von einem generativen Modell generierte synthetische Vibrationsdaten sind effektiv bei der Verbesserung der Klassifikationsleistung eines Klassifikationsmodells. Dies demonstriert die Fähigkeit eines generativen Modells, einige der Experimente zu ersetzen, und wird erwartet, dass es zu Zeit- und Kosteneinsparungen beiträgt. Bei beiden, generativen und Klassifikationsmodellen, ist die Optimierung der Hyperparameter kritisch für ihre Implementierung.

Zusammenfassend wird erwartet, dass die Ergebnisse dieser Studie die Abhängigkeit von Feldexperimenten minimieren und das physische Generieren und digitale Erweitern von Vibrationsdaten im Labor schnell und einfach ermöglichen sowie hochwertige Daten für datengesteuerte Ansätze bereitstellen, während sie im Vergleich zu Feldexperimenten Zeit und Geld sparen. Basierend auf der Flexibilität, Geschwindigkeit und den niedrigen Kosten des FDM 3D-Drucks, als praktische Testplattform, kann das Maßstabsmodell flexibel für die Defekterkennung oder andere Anwendungen für Fahrzeuge und Infrastruktur skaliert werden. Dieses skalierbare Modell wird erwartet, den Prozess des schnellen Testens und Verfeinerns von Ideen in der vorläufigen Forschungsphase erheblich zu beschleunigen und die Forschung zur Optimierung des Eisenbahnbetriebs und der Wartung auch innerhalb der Einschränkungen von Feldexperimenten zu verbessern.

1 Introduction

In the railway industry, continuous research efforts and the pursuit of technological advancements play an important role in enhancing the overall safety, reliability, and cost-effectiveness of railway operations, as well as in ensuring on-time departures and arrivals to improve service punctuality. To keep up with ever-changing technological trends and requirements, the railway industry requires a variety of field experiments covering different aspects of railway operations, vehicles, and infrastructure. However, conducting field experiments with real vehicles and tracks is often difficult due to several reasons: the complexity of train scheduling, safety concerns, budget constraints, difficulties in changing vehicles and infrastructure, and operational disruptions.

As an alternative to these difficulties in field experiments, various types of scale models have been developed and used in laboratories for different research purposes in the preliminary research phase (Naeimi, Li, et al. 2018). Among the various possible applications of railway scale models, this study utilised a railway scale model to measure vibration data from the axle box adjacent to wheel flats, and then it incorporated a generative model to improve the data-driven classification of wheel flats. Based on these results, the optimal geometry for wheel flats were determined for follow-up studies. Figure 1 shows a simulated wheel flat, which is a common type of defect found on the wheels of freight vehicles and high-speed vehicles. Wheel flats typically occur due to plastic deformation of the wheel tread during skidding on a rail after sudden braking. A discontinuity between the defective wheel and the rail causes impulsive vibrations.

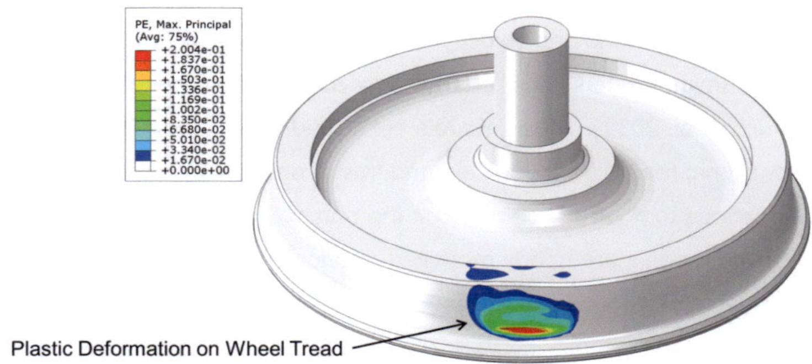

Figure 1:   **Example of Simulated Wheel Flat in Scale Model using Finite Element Method (FEM)**

1.1 Problem Statement

Both the railway industry and academic institutions have been actively involved in the research and development of diagnostic systems for railway vehicles and infrastructure. Unlike manufacturers and operators in the railway industry, it is relatively difficult for research groups at academic institutions to obtain high-quality data from field experiments with real railway vehicles and tracks. They tend to use data from numerical simulations rather than field experiment as an alternative. This makes it difficult for them to obtain high-quality data for their initial studies, especially in the preliminary research phase when planning a new research project or before it starts in earnest. In this phase, initial ideas need to be conceptualised and refined through trial and error, and the lack of good quality data hinders this process.

In addition, there has been a significant shift towards data-driven approaches in the development of diagnostic systems. Despite the increasing demand for high-quality data and the difficulties in obtaining it efficiently and economically, these data-driven diagnostic systems are becoming increasingly preferred (Xie, et al. 2020). This approach performs exceptionally well at self-learning feature extraction and defect detection, in contrast to traditional discriminative algorithms that require extensive user involvement for manual feature extraction and rule-based decision making (Krizhevsky, Sutskever and Hinton 2012).

The characteristics of high-quality data can be described by the 3V (volume, variety, and velocity) that are often used in 'Big Data' literature, where volume refers to the quantity of data, variety to the diversity of data, and velocity to the speed at which data are collected and processed (LaneyD. 2001). In this sense, relying solely on field experiments may lack volume and velocity, even if variety is sufficient. As an alternative, numerical dynamics models have been widely used to compensate for volume and velocity (Bruni, et al. 2011). These models are based on the common assumption that they closely mimic real railway vehicles and tracks and are simplified based on additional assumptions that vary depending on the purpose of the application. However, if these dynamics models are poorly modelled or uncalibrated, they may quickly generate large amounts of data, which are unfortunately of poor quality, thereby leading to ineffective diagnostic systems.

Conversely, if a railway scale model is constructed to closely approximate a real railway vehicle and track, it can provide an advantage over numerical dynamics models. This model ensures that the generated data naturally adhere to correct assumptions and accurately reflects the laws of physics. Although railway scale models also require certain assumptions and estimates, they inherently follow the laws of nature and more closely align with reality, unlike numerical dynamics models, which depend on complex theoretical equations, many assumptions, and estimates.

However, the process of downscaling a real railway vehicle and track to a scale model does not perfectly downscale all physical variables due to the limitations imposed by the laws of similarity (Jaschinski, et al. 1999). Specific law of similarity is only meaningful for certain physical variables depending on the application, which necessitates a careful application within the effective ranges of these laws. Ideally, if materials, manufacturing methods, and structures remain unchanged before and after downscaling, a scale model should align quantitatively with a real target system as dictated by the law of similarity. Nevertheless, achieving perfect similarity across all physical variables is usually difficult due to inevitable design compromises and simplifications, and successful downscaling is limited to specific physical variables. Furthermore, it becomes more difficult to achieve high similarity when different materials, manufacturing methods, or structures need to be introduced into a scale model. And although less demanding than field experiments, collecting data from a scale model in a laboratory still requires significant time and effort. The key problems are summarised as follows:

- **Difficulty in Obtaining High-Quality Data from Field Experiments**: At the preliminary research phase, it is not easy to obtain high-quality data from field experiments, which is essential for data-driven studies in the laboratory.
- **Limitation of Numerical Dynamics models**: It has been widely used as a practical alternative, but if it is modelled based on incorrect assumptions or poorly calibrated, a defect detection model trained on its data will be useless.
- **Limitation of Scale Models**: Compared to numerical dynamics models, scale models inherently follow the laws of nature better, but they may not be accurate downscaled and may show different dynamic and vibration characteristics due to the downscaling constraints. Dynamic testing in the laboratory still needs considerable time and effort compared to numerical approaches.

1.2 Scope and Objectives

The research framework was designed to include both downscaling and upscaling processes between a full-scale model in the field and a scale model in the laboratory, as shown in Figure 2. The scope of this study was narrowed down to research activities in the laboratory with three key questions: 1) could a scale model, constructed using Fused Deposition Modelling (FDM) 3D-printers and PLA filaments, physically generate high-quality vibration data required for studying the data-driven detection of wheel flats?, 2) is it effective to augment vibration data using a generative model in terms of data volume and velocity?, and 3) how could a 3D wheel flat be designed to be equivalent or similar to the 2D wheel flat models widely used in railway vehicle-track dynamics models?. The main objective is to facilitate preliminary studies on data-driven defect detection, which requires high-quality vibration data in a laboratory environment, based on a scale model and a generative model. The key tasks are as follows:

- Design and construct a scale model using FDM 3D-printing and PLA filaments
- Measure raw vibration data through dynamic testing on a short straight track
- Build a LSTM-based generative model for producing synthetic vibration data
- Build a CNN-based classification model for identifying wheel flat geometries
- Optimise the hyperparameters of the LSTM network using a genetic algorithm and those of the CNN model using a grid search algorithm, respectively

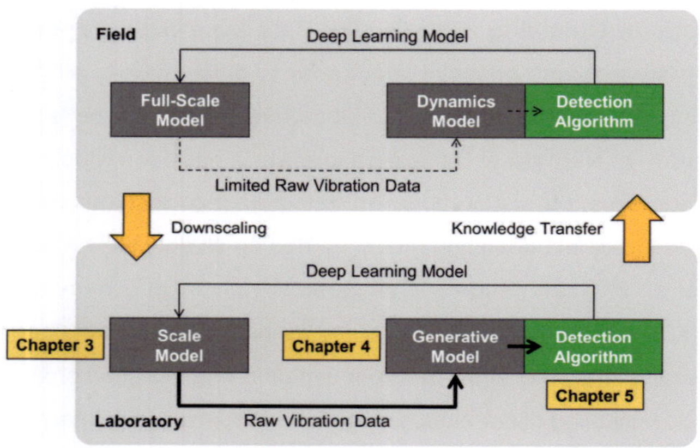

Figure 2:   **Research Framework for Defect Detection based on Augmented Vibration Data**

1.3   Outline

Chapter 2 follows the problem statement introduced in Chapter 1 with a literature review on the modelling and detection of wheel flats to explore research trends, technical issues, and research gaps in more detail.

Chapter 3 describes the process of design, construction, and subsequent vibration analysis of a 3D-printed railway scale model with wheel flats, which is designed to specifically capture raw vibration data excited by wheel flats and amplified by the bending modes of a flexible wheelset. Three modelling methods are proposed to investigate the influence of wheel flat geometries on vibration data and classification performance, and to determine which modelling method is most appropriate. Finally, dynamic testing is performed on a short straight track to measure raw vibration data from the axle box positioned adjacent to wheel flats.

Chapter 4 explores an LSTM-based generative model for producing synthetic vibration data, which serves as an alternative to the considerable time and effort required to measure raw vibration data. For better data augmentation, raw vibration data is segmented and filtered by pre-processing, and hyperparameters are optimised using a genetic algorithm. Synthetic vibration data are validated by performing a correlation analysis with raw vibration data using the Modal Assurance Criterion (MAC) in terms of cosine similarity.

Chapter 5 focuses on identifying the occurrence and types of wheel flats using a CNN-based classification model. The process begins with the preparation of specific datasets for case studies. The architecture of a CNN model is then described, and the optimal hyperparameters are determined using a grid search algorithm. The performance of the classification models, trained using different data augmentation strategies, is compared. The chapter concludes with a discussion on the effects of hyperparameters, data augmentation, and the modelling method on classification performance. It then proposes an effective wheel flat modelling method and data augmentation strategy for follow-up studies.

Chapter 6 concludes this dissertation with a summary of the main findings and discusses future work to further extend the contribution to the railway field.

2 Review and Identification of Research Gaps in Modelling and Detection of Wheel Flats using Scale Models

This chapter presents a literature review focusing on the keyword 'Wheel Flat' in relation to modelling methods, implementation in scale models, and producing synthetic vibration data via generative models, defect detection methodologies. This review has identified research trends, technical issues, and research gaps in existing studies. Based on these findings, the subsequent chapters (3, 4, and 5) are structured to make contributions toward addressing these identified gaps.

2.1 Detection Methodologies of Wheel Flats

This subchapter focuses primarily on existing studies that use vibration data for the detection of wheel flats. It also includes a related series of studies that identify track defects on a 1/87 scale model, which was constructed in the form of a vehicle on an oval track by the Institute of Railway and Transportation Engineering (IEV) at the University of Stuttgart (Rapp, et al. 2019). Existing studies are categorised into traditional and deep learning-based methods, as summarised in Table 1.

Regardless of the categorisation in Table 1, existing studies have been obtained vibration data from Multi-body Dynamics (MBD) models and test rigs in the laboratory, as well as from real vehicles and tracks in the field. However, measuring vibration data from real vehicles and tracks is not always feasible. As an alternative, a few studies used MBD models and test rigs together. Based on the assumption that there is a high similarity between real and simulated vibration data, training and validation were performed on simulated vibration data, while testing was conducted on a limited amount of real vibration data (Bosso, Gugliotta and Zampieri 2018), (Li, Liu and Wang 2016).

Vibration data, measured in units of acceleration or force, comes from two primary sources. For vehicles, they come from accelerometers mounted on the axle box and at the centre of carbody. For tracks, they are collected from strain gauges placed on the side of the rail, as well as from accelerometers and ultrasonic sensors positioned on the sleepers. Since wheel flats and track defects predominantly generate defect-induced energy in the vertical direction, existing studies typically use vertical acceleration and force as the key inputs for feature extraction and detection algorithms.

Category	References
Traditional Methods	Simulation, MBD, Wheel Flat, Wayside, Vertical Force/Acceleration - CWT/PCA (Mosleh, et al. 2023) Simulation, MBD, Wheel Flat, Wayside, Vertical Acceleration - Kurtosis (Mosleh, et al. 2021) Simulation, MBD, Wheel Flat, Wayside, Vertical Force - Envelope Spectrum (Mosleh, et al. 2020) Simulation + Experiment, Wheel Flat, Axle Box, Vertical Acceleration - RMS (Bosso, Gugliotta and Zampieri 2018) Simulation + Experiment, Wheel Flat, Axle Box, Vertical Acceleration - EMD (Li, Liu and Wang 2016) Experiment, Test Rig, Wheel Flat, Wayside, Ultrasound - Bandpass Filter (Brizuela, Fritsch and Ibáñez 2011) Simulation, MBD, Wheel Flat, Body, Vertical Acceleration - WD (Jia and Dhanasekar 2007)
Deep Learning -based Methods	Experiment, Train, Wheel Flat, Axle Box, Vertical Acceleration - STFT + CNN (Ye, et al. 2023) Simulation, MBD, Wheel Flat, Axle Box, Vertical Acceleration - STFT + CNN (Shim, et al. 2022) Simulation, MBD, Wheel Flat, Axle Box, Vertical Acceleration - FFT + CNN, LSTM (Sresakoolchai and Kaewunruen 2021) Experiment, Train, Wheel Flat, Axle Box, Vertical Acceleration - STFT/CWT/HT/WPT + Lightweight CNN (Shi, Ye, et al. 2021) Simulation, MBD, Wheel Flat, Axle Box, Vertical Acceleration - Wavelet + ResNet (Chen, et al. 2021) Experiment, Scale Model, Track Defects, Vertical Acceleration - MA Filter + CNN (Bahamon-Blanco, Liu and Martin 2021) - MA Filter + LSTM (Bahamon-Blanco, et al. 2020) - MA Filter + Bagged Decision Tree (Bahamon-Blanco, et al. 2019) Experiment, Wayside, Wheel Flat, Vertical Force - Wavelet + SVM/CNN (Krummenacher, et al. 2017)

Table 1:   **List of Wheel Flat Detection Methods using Vibration Data**

Traditional methods, including early machine learning models such as Support Vector Machines (SVMs) and Bagged Decision Tree, utilise statistical and signal processing techniques to manually extract features from raw vibration data. These features are then evaluated according to various standards and criteria. In contrast, deep learning-based methods automatically extract features in different ways de-pending on the type of deep learning models. In supervised learning models, these extracted features are linked to labels and then classified through a fully connected layer. Although deep learning-based approaches are becoming increasingly popular due to recent technology trends, many existing studies still explore traditional methods for feature extraction because of the transparent nature and simplicity of the internal calculation process.

Wheel flats and track defects typically produce impulsive signals, which are shaped by structural resonances and damping ratios. The selection of appropriate statistical and signal processing techniques is important for interpreting vibration data. Under-standing the shape of the vibration data is necessary for effective feature extraction. For instance, employing a long-time block in time-frequency analysis, such the Short-time Fourier Transform (STFT), may lead to the blurring of transient features.

Within the implementation of traditional methods, several statistical techniques are used to analyse vibration data. Kurtosis measures the deviation from a normal distribution. The Root Mean Square (RMS) is utilised to quantify the energy contained in vibration data. The Principal Component Analysis (PCA) is employed to identify the primary axes of the dataset, which helps in reducing its dimensionality while preserving the most significant features. In terms of signal processing techniques, The Continuous Wavelet Transform (CWT) and Wavelet Decomposition (WD) are often preferred over Fourier transforms for their ability to capture transient features in signals. The Envelope Spectrum is useful for tracking amplitude modulations. A Bandpass Filter is used to allow only a specific frequency band to pass through. Lastly, The Empirical Mode De-composition (EMD) is applied to extract complex, non-periodic transient features.

Machine learning models, such as SVMs and Bagged Decision Tree, do not have a neural network structure for feature extraction, so they also require pre-processing to extract features, just like traditional methods. On the other hand, deep learning models, such as Convolutional Neural Networks (CNNs), ResNets (Residual Networks), and

Long Short-Term Memory (LSTM) networks, typically consist of a neural network structure for feature extraction in the spatial or temporal domain and a classifier to link the extracted features with labels. Thus, deep learning model-based approaches can classify vibration data with high accuracy compared to traditional methods.

However, compared to traditional methods, deep learning approaches generally require a significant amount of vibration data to effectively train a neural network structure, which complicates dataset preparation than before. In addition, they often require high computational load and involve iterative optimisation of hyperparameters, which can vary across different deep learning models. This has led to the development of lightweight models that can run on a low-cost embedded system (Shi, Ye, et al. 2021). In conclusion, deep learning models are more effective in classifying them, but some technical difficulties still justify the need for continuous studies on traditional methods. From the literature, the following points are noted:

- **Sources for Vibration Data**: Existing studies on defect detection have been relying on vibration data from MBD models, test rigs, and field experiments. Due to the difficulty of measuring real vibration data, numerical approaches are often used instead. Since the vibration generation mechanism is dominated in the vertical direction, vertical accelerations and forces were commonly used as inputs in studies for the detection of wheel flats and track defects.
- **Evolution from Traditional Methods to Deep Learning Models**: Traditional methodologies that manually extract features from raw data using signal processing techniques have evolved into deep learning models that automatically extract them through training and improve classification performance.
- **Trade-offs Between Traditional Methods and Deep Learning Models**: Despite their high classification accuracy and automated feature extraction capabilities, deep learning models require large amounts of data and high computational resources for the training process. This is why lightweight deep learning models are being developed and traditional methods are still widely used.
- **Approaches Aligned with Vibration Data Characteristics**: Effective classification depends on properly understanding the characteristics of vibration data. Whether using traditional methods or deep learning models, selecting the right

one is crucial to align best with the vibration data. This is essential for accurately extracting features and enhancing classification performance.

2.2 Implementation of Wheel Flats in Scale Models

Various forms of scale models have been developed for different research purposes in sizes that can be used in laboratories. Table 2 lists the types, scales, and purposes of scale models used for vehicle dynamics applications. In the categories 'Vehicle on Track' and 'Bogie on Roller Jig', the existing studies are generally related to vehicle dynamic behaviour such as pitch, roll, yaw, and bounce, which are global modes in the low frequency range. However, there are fewer studies dealing with local modes of defective structural parts in the higher frequency range. For wheel flats, apart from studies on a 1/5 scale roller jig (Liang, et al. 2013), there is no other example. In the categories 'Wheel on Ring Rail' and 'Twin Discs', contact and wear between wheel and rail have been mainly studied. From the literature, the following point is identified:

- **Lack of Scale Model**: There is no scale model in the category 'Vehicles on Track' yet that can implement wheel flat-induced vibration from axle box accelerometers. Moreover, most existing scale models have been constructed from metal, such as steel and aluminium, using traditional machining methods.

Category	References
Vehicle or Bogie on Track	1/10, Track Irregularities (Chamorro, et al. 2022) 1/87, Track Defect (Rapp, et al. 2019) 1/5, Lateral Instability (Lu and Yang 2019) 1/5, Lateral Instability (Kim, Park and You 2008) 1/5, Lateral Instability (Michitsuji and Suda 2006) 1/10, Derailment (Hung, et al. 2010)
Bogie on Roller Rig	1/5, Wheel Flat (Liang, et al. 2013) 1/5, Independent Wheel Control (Kurzeck and Valente 2011) 1/5, Ride Quality (Shin, et al. 2014) 1/5, Hardware-in-the-loop System (Oh, et al. 2019)
Wheel on Ring Rail	1/5, Wear, Wheel-Rail Contact (Naeimi, Li, et al. 2017) 1/5, Wear, Wheel-Rail Contact (Zhu, Thompson and Jones 2011)
Twin Discs	1/2, Wear, Wheel-Rail Contact (Vuong, et al. 2011) 1/2, Wear, Wheel-Rail Contact (Takikawa and Iriya 2008)

Table 2: **List of Scale Models for Applications in Mechanical and Control System**

2.3   Modelling Methods for Wheel Flats

Table 3 lists the modelling methods of wheel flats. Many existing studies on modelling wheel flats have mainly focused on improving 2D wheel flat models, defined as a function of wheel radius, for calibrating railway vehicle-track dynamics models (Uzzal, Ahmed and Bhat 2013). Figure 3 shows a typical example of wheel flat with sharp edges. More realistic 3D wheel flats have also been modelled by applying analytical 3D wheel-rail contact geometries as input to an existing 2D wheel-rail contact model, defined by a set of springs based on Hertzian contact theory (Zunsong 2019), (Pieringer, Kropp and Nielsen 2014). In rare cases, some studies have used experimental vibration data, containing impulses of wheel flats, from a real vehicle and track or a scale model. In a real freight vehicle, wheel flats were artificially implemented by machining and emergency braking, respectively (Shi, et al.2022). In a scale model of a roller test rig, wheel flats were also artificially implemented by machining (Liang, et al. 2013). The modelling process of wheel flats was not described in detail compared to the existing studies numerically carried out with dynamics models. Unfortunately, it is not known in which 3D geometry the wheel flats were implemented on the wheel tread. The only information available is the approximate size and position of the wheel flats on the wheel tread. The key point of this subchapter is summarised as below:

- **Lack of 3D Wheel Flat**: Except for some pictures of real 3D wheel flats and the 3D wheel flat geometries used as input to the wheel-rail contact model, there is no reference for modelling the 3D geometry of wheel flats on a 3D-printed wheel.

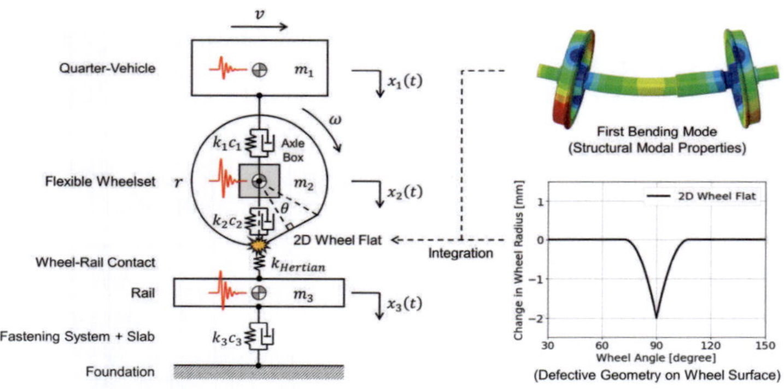

Figure 3:　　　　**Simple Vehicle-Track Dynamics Model with Wheel Flat**

Category	References
Dynamics Model	2D Wheel Flat (Uzzal, Ahmed and Bhat 2013) 2D Wheel Flat + 3D Wheel-Rail Contact (Zunsong 2019) 2D Wheel Flat + 3D Wheel-Rail Contact (Pieringer, Kropp and Nielsen 2014)
Freight Wagon	3D Wheel Flat by Machining and Emergency Brake (Shi, Ye, et al. 2022) 3D Wheel Flat by Machining (Bosso, Gugliotta and Zampieri 2018)
Scale Model	3D Wheel Flat by Machining (Liang, et al. 2013)

Table 3:　　　　**List of Modelling Methods for Wheel Flats**

Category	References
Generative Adversarial Networks (GAN)	Wheel Defect in Freight Wagon (Shi, et al.2022) Bearing Defect in Test Rig (Timo, et al. 2023) Bearing Defect from Open-Source Datasets (Guo, et al. 2022) Wear in Milling Machine (Shah, et al. 2022) Gear Defect in Test Rig (Ma, et al. 2021)
Variational Autoencoder (VAE)	Fastener Damage in Track (Wang, et al. 2023) Structure Damage in Bridge (Ma, et al. 2020)
GAN+VAE	Bearing Defect in Electrical Locomotive (Liu, et al. 2022) Hunting Instability in High-Speed Train (Ning, et al. 2022) Bearing Defect in Test Rig (Liu, et al. 2023) Bearing Defect in Test Rig (Singh and Harsha 2023) Bearing Defect in Test Rig (Liu, et al. 2021)
Long Short-Term Memory (LSTM)	Noise in Vehicle Steering Gear (Bu, Moon and Cho 2021) Activity Recognition in Excavator (Rashid and Louis 2019)

Table 4:   **List of Generative Models for producing Synthetic Vibration Data**

2.4   Synthetic Vibration Data Generation via Generative Models

Table 4 lists the generative models used in related studies and their purposes. Based on the literature review, the following three generative models were mainly compared.

Generative Adversarial Networks (GANs) are widely used due to their ability to generate high-resolution synthetic data. GANs achieve this through a competitive training process between generator and discriminator. Variational Autoencoders (VAEs), known for their efficiency in encoding data into a compressed latent space, are often used in combination with GANs. This combination leverages the strength of VAEs in feature extraction and the strength of GANs to generate high-quality synthetic data, thereby enhancing both feature extraction and training stability over using VAEs alone. LSTM networks excel in handling sequential data, effectively capturing long-term dependencies in time-series data. However, they are relatively less used as a generative model compared to GANs and VAEs.

The primary purpose of using generative models to generate synthetic vibration data is to augment experimental vibration data at the component level, such as bearings and gears, typically measured on test rigs in laboratories. Few studies have directly conducted on vehicles and tracks at the system level, not the component level. One of the existing studies created four types of wheel flats on a real freight vehicle through grinding and emergency braking, then measured vibration data from the axle box, and attempted to augment it with various methods including GANs (Shi, et al.2022). The experimental vibration data contains many impulses induced by wheel flats. Synthetic vibration data augmented using GANs did not contribute to classification performance compared to traditional dynamics model-based approaches. High computational load in training and hyperparameter optimisation was also mentioned as a difficulty during implementation. The key points of this subchapter are summarised as follows:

- **Board Applications of GANs and VAEs**: GANs and VAEs are widely employed for generating synthetic vibration data in related studies because of their robust data generation capabilities. Existing studies tend to use data measured on test rigs in laboratory rather than field experiments. As a result, there are more examples at the component level (e.g. bearings and gears) rather than at the system level (e.g. vehicles and tracks).

- **Combination of Different Deep Learning Models**: By utilising different deep learning models together, it is possible to combine the strengths of each architecture. This can improve the quality of the synthetic data and the stability of the training process. In Table 4, 'GAN+VAE' means a combination of GANs, which utilise an advanced training mechanism, and VAEs, known for efficient data encoding. This is more effective than either alone.

- **Limitations of GANs in Data Augmentation**: GANs are a popular choice for data augmentation because of their advanced learning mechanism, but their effectiveness can vary based on the characteristics of the vibration data and the optimisation of hyperparameters. They may not always generate high-quality synthetic vibration data, which can lead to poorer classification results compared to using vibration data obtained from well-calibrated dynamics models.

- **Rare Applications of LSTM Networks in Data Augmentation**: LSTM networks, despite their proficiency in processing time-series data, are less frequently used for generative modelling compared to GANs and VAEs.

2.5   Connecting Research Gaps to Current Study

This study fundamentally started from the question, "Is there an alternative way to physically implement abnormal vibrations caused by wheel flats in the laboratory instead of field experiments?" In the previous subchapters, existing studies were reviewed to identify possible alternatives and research gaps. The detailed research direction is illustrated in Figure 4.

If only the wheel flat was considered, the 'Bogie on Roller Rig' type as already used in the existing study (Liang, et al. 2013) would probably be sufficient, but the 'Vehicle on Track' type was chosen instead in order to include inertial effects in terms of vehicle dynamics and in view of follow-up studies for other purposes. The 3D geometry of wheel flats was considered for the 'Bogie on Roller Rig' type, but there is no detailed description of how its 3D geometry was modelled. There is also no example in the literature of implementing wheel flats on the 'Vehicle on Track' type. All the scale models found through the literature review had in common that they were constructed using metal materials and traditional machining, and there were no scale models that used 3D-printing technology to fabricate almost all the structural parts. This type of scale model, which has never been constructed before, and 3D-printing technology aims to physically contribute to reducing the difficulties in obtaining high-quality vibration data as well as to increasing flexibility in design and construction.

Once the raw vibration data is obtained from the 3D-printed scale model, it is important to implement preprocessing, a generative model, a classification model, and hyperparameter optimisation as lightweight as possible based on an understanding of the characteristics of the vibration data so that it makes possible to minimise the requirements for data quality and computational load while taking advantages of deep learning models. For the generative model, an LSTM network was chosen due to its ease of handling sequence data. If there are difficulties in training, it will be combined with GANs. Since the classification model uses the averaged spectra as input, a CNN model was chosen because of its robustness to classification. While existing studies have mainly relied on combinations within field experiment, test rig, and simulation to improve data quality, the generative model aims to improve data quality without additional experiment or simulation. The classification model aims to relatively compare the different

3D geometries of wheel flats in terms of distinct classification and to determine which one is more suitable for subsequent size classification studies.

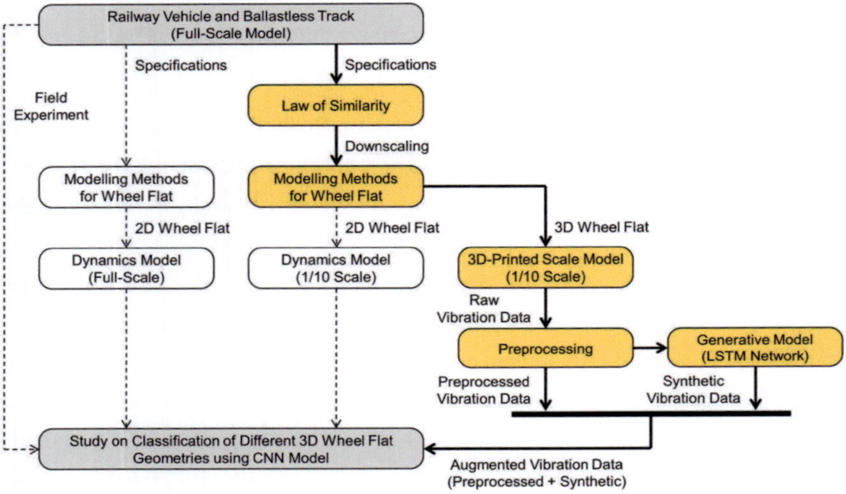

Figure 4: **Research Direction established from Literature Review**

3 Design, Construction, and Vibration Analysis of 3D-Printed Scale Model with Railway Wheel Flats

This chapter describes how the 3D-printed scale model was designed and constructed with reference to the real metro vehicle and ballastless track. In order to obtain the raw vibration data from the axle box positioned adjacent to wheel flats, the following tasks were carried out: 1) designing and constructing the scale model using FDM 3D-printing, 2) designing in detail the flexible wheelset and the wheel flat geometry, 3) developing the measurement and control system, and 4) conducting dynamic testing.

3.1 Design of Scale Model

The scale model was designed based on a railway vehicle-track dynamics model that has been widely used for various research purposes, rather than simply trying to downscale a real metro vehicle and ballastless track as they appear. A 1/10 scale was chosen because of the availability of laboratory space for dynamic testing, budget, materials, machining methods, and connectivity with follow-up studies. Due to constraints and inevitable compromises during the design and construction process, it is difficult to perfectly downscale a real metro vehicle and track to a scale model while maintaining all the structural characteristics. Therefore, the scope of application of a scale model needs to be clearly defined. In this study, it is important to ensure that the bending modes of the wheelset and the influence of wheel flats are clearly captured qualitatively in the raw vibration data measured using accelerometers at the axle box. These dynamic characteristics were prioritised so that the design of the scale model could be designed with a high degree of similarity to a real metro vehicle and track.

Figure 5 shows a railway vehicle-track dynamics model. The vehicle consists of two bogie models, which are connected to each other by primary and secondary suspensions, including several spring and damper elements. The ballastless track rests on a rigid foundation. Cement and emulsified asphalt (CA) mortar, concrete base and slab are laid in a sequence. The rails are connected to the slab using a fastening system, including spring and damper elements. The vehicle and ballastless track are connected using a wheel-rail contact model based on the Hertz contact theory. A wheel flat is applied to a wheel as a function of the wheel radius. In dynamics modelling, due to the high contribution of the wheelset flexibility to the raw vibration data from the axle box accelerometers, it is often considered as the only finite element model, unlike other bodies which are assumed to be rigid. Although the scale model cannot be designed in the same dichotomous way as a dynamics model, with bodies divided into rigid and flexible parts, it was designed with the intention of bringing a numerical dynamics model created for the study of wheel flats into physical form in the laboratory. In particular, the flexibility of the wheelset and the geometry of the wheel flats were considered important.

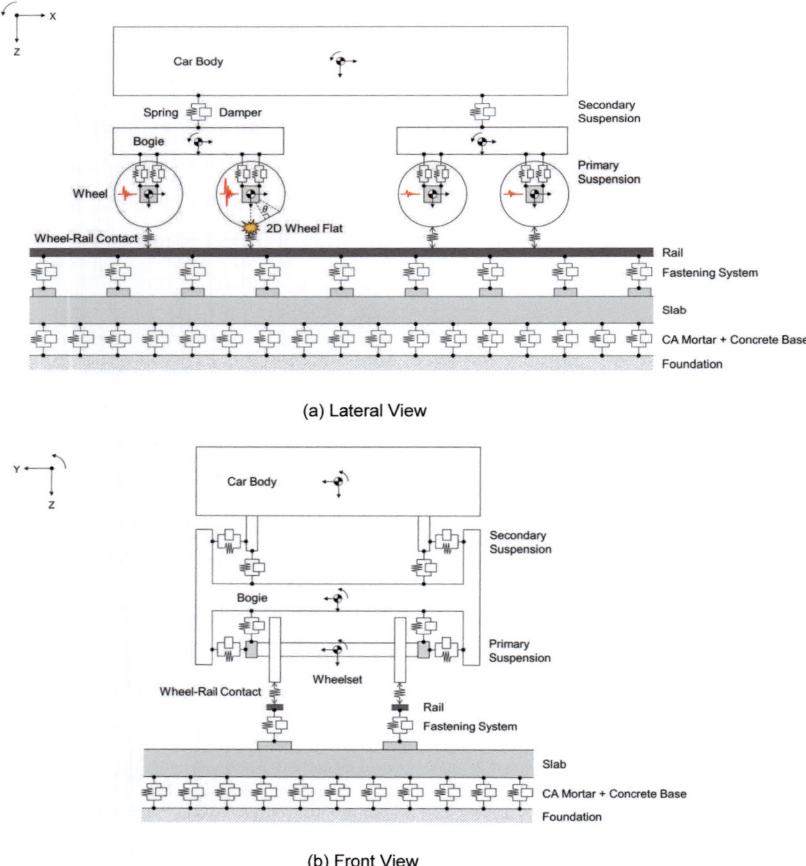

(a) Lateral View

(b) Front View

Figure 5:   **Railway Vehicle-Track Dynamics Model with Wheel Flat**

Category	Parameter	Full-Scale	1/10-Scale
Car Body	Mass	19,000 kg	19 kg
	Length X	19.52 m	1.952 m
	Length Y	2.88 m	0.288 m
	Length Z	3.8 m	0.38 m
Bogie	Mass	6,500 kg	6.5 kg
	Wheelbase	2.3 m	0.23 m
	Distance between Bogies	12.6 m	1.26 m
Wheelset	Mass	1,138.8 kg	1.1388 kg
	Wheel Diameter	0.84 m	0.084 m
Axle Box	Mass	110 kg	0.11 kg
Primary Suspension	Stiffness Z	1,300,000 N/m	1,300 N/m
	Damping Coefficient Z	1,840 Ns/m	1.84 Ns/m
Secondary Suspension	Stiffness Z	424,000 N/m	424 N/m
	Damping Coefficient Z	25,000 Ns/m	25 Ns/m
Operating Condition	Speed	0 to 80 km/h (0 to 22.2 m/s)	0 to 8 km/h (0 to 2.2 m/s)

Table 5: List of Original and Downscaled Design Parameters

Table 5 shows the original and downscaled design parameters provided by Hefei Metro. The parameters of the full-scale model were downscaled according to the law of similarity (Appendix I: Downscaling Strategies by Research Objective), which ensures that there is no change in time and frequency after downscaling, which is often used for the 'Bogie on Roller Rig'. Based on this information, it is possible to describe the vehicle dynamic behaviour as global modes of the vehicle in the low frequency range below 50 Hz, but there is insufficient information to describe the resonance modes of the wheelset as local modes in the higher frequency range. This problem has been solved by estimating the required information in reverse by referring to the resonance frequencies of the wheelset, their mode orders, material properties, etc. published in related studies dealing with similar metro vehicles and tracks (Cai, et al. 2019).

3.1.1 Vehicle and Track

Based on the downscaled design parameters in Table 5, the vehicles and ballastless track were designed as closely as possible to the real metro vehicle and track using Autodesk Fusion 360. As the vehicle consists of two bogies that can operate independently on the track, the details are introduced in detail at the bogie level rather than at the vehicle level. Figure 6 shows the detailed design of the bogie model and track model in isometric and side views. Unlike many studies that construct scale models using traditional machining methods with materials such as steel or aluminium alloys, this study chose to use FDM 3D-printers and Polylactic acid (PLA) filaments. Therefore, it was crucial to design a shape that would print with high quality, considering the material properties, the printing mechanism, and the size of 3D-printer's bed plate.

Most of the structural components for the vehicle part were printed in the laboratory, except for some components under high loads, such as the bogie centre frame made of 6063-T5 aluminium profile. The drivetrain has Brushless Direct Current (BLDC) motors at the front and rear. The motor torque is transmitted to the wheels via the gearbox with a 5:1 gear ratio. The primary and secondary suspension was implemented in more detail than scale models in related studies, using springs and dampers based on the degrees of freedom of the dynamics model. On top of the bogie frame, the control box contains the electrical and electronic components for measurement and control: low-cost embedded systems (Jetson Nano and Arduino R1), batteries, sensors, etc. Four IMUs were installed in the axle box to measure the 3-axis axle box accelerations and are connected to Arduino R1 via Inter-Integrated Circuit (I^2C) interface.

A short straight track of 2.4 m in length was also printed for dynamic testing. The rails have an inclination of 1/40 and a track gauge of 0.1435 m. In order to control the vertical stiffness of the slab while minimising printing time and filament quantity, the slab was printed without any surface layers and with only a Gyroid infill structure of 20% density. Since the maximum printable size of the bed plate is 300 x 300 mm, the rails and slabs were printed in 297.5 mm lengths. Therefore, the track irregularities are expected to be high. The rubber pads between the rails and the slab were printed using Thermoplastic Polyurethane (TPU) filament (like rubber). In the dynamics model of the track, the slab is modelled as if it is floating on a rigid foundation, so the track was placed on top of the office floor mats without being fixed.

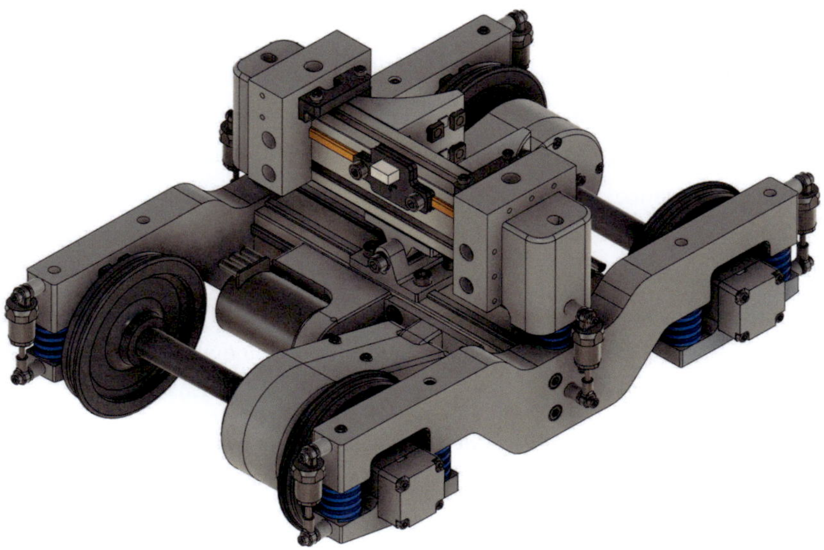

(a) Isometric View

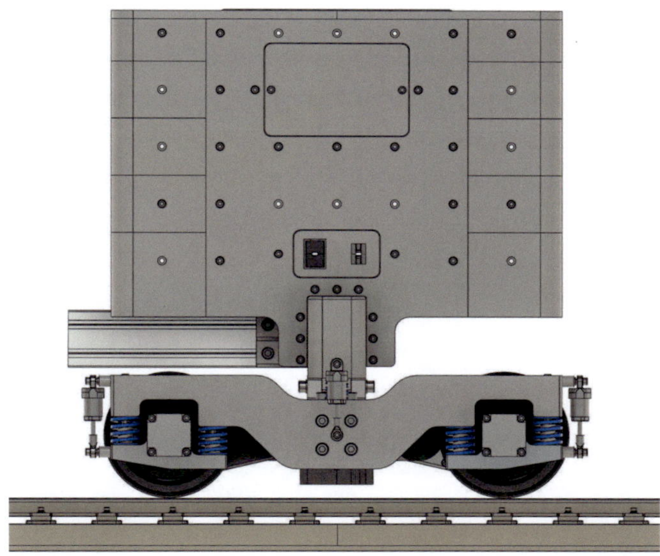

(b) Lateral View

Figure 6:   **Detailed Design of 1/10 Scale Bogie Model on Ballastless Track**

3.1.2   Flexible Wheelset

If a railway vehicle-track dynamics model with primary and secondary suspensions, including springs and dampers, has only rigid wheelsets, the vibration responses observable at the axle box or the carbody are confined to roll, pitch, yaw, and bounce. In this case, the influence of the wheelset's vibration modes, especially bending modes, is not implemented in the measured vibration data. This means that the impulsive vibration energy generated by a discontinuous wheel-rail contact due to wheel flats is not affected by the structural resonances of the wheelset as it flows through the wheelset to the bogie and carbody. Therefore, the flexibility of the wheelset is important in order to properly implement the vibration response caused by wheel flats.

Figure 7 shows the detailed design of a wheelset that allows the wheel conditions to be changed during dynamic testing. The axle features a hexagonal bore for torque transmission and to minimise slippage. Wheels are attached to the axle with M3 threaded inserts and secured with M3 screws, simplifying the process of applying wheel defects for dynamic testing. The wheel conforms to the TB/T 449 standard, ensuring that it maintains the correct geometry for accurate experiments, despite its reduced size. A GT2 profile timing belt pulley is connected to a BLDC motor in the gearbox with a 4:1 gear ratio. The wheel and axle use a Gyroid infill structure of 20% density, chosen for its isotropic mechanical properties and high 3D-printing speed.

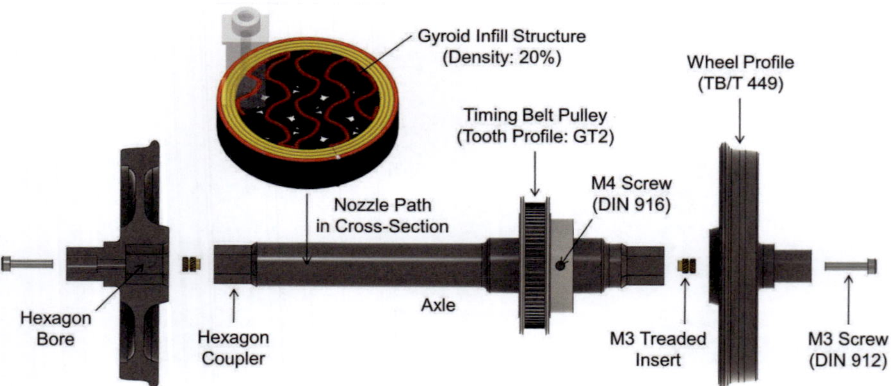

Figure 7:   **Detailed Design of Downscaled Wheelset**

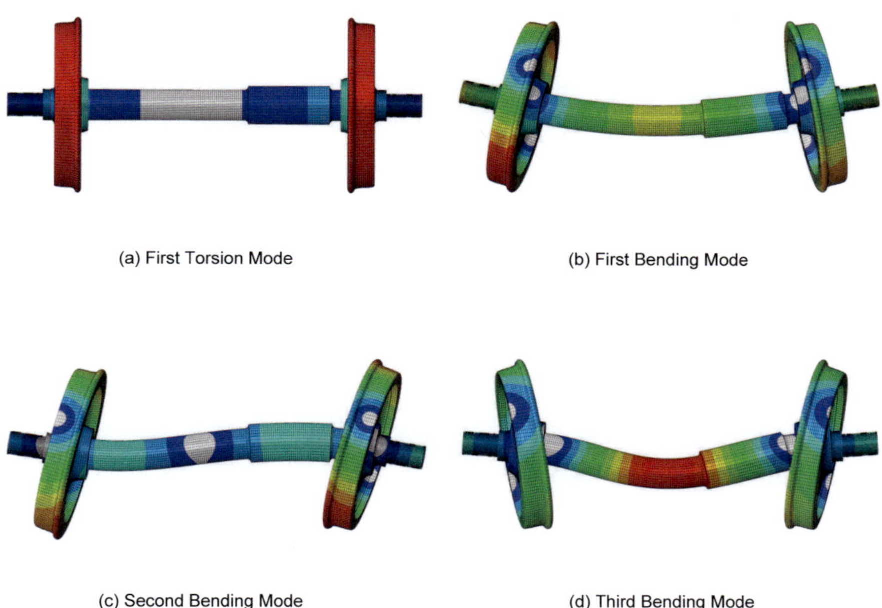

(a) First Torsion Mode

(b) First Bending Mode

(c) Second Bending Mode

(d) Third Bending Mode

Figure 8: **Mode Shapes of Flexible Wheelset**

For an accurate representation of vehicle-track dynamics in the scale model, the wheelset should have proper structural resonance characteristics, as shown in Figure 8. The mode shapes are essential because they can amplify the impulsive vibration energy resulting from wheel flats. Ignoring these structural resonance characteristics could lead to a significant underestimation of their influence on the vibration data.

Therefore, it is important that the effects of these important mode shapes, especially bending modes, can be adequately captured within the measurable frequency range without being distorted by the limitations of low-cost measurement systems. The maximum measurable frequency is the Nyquist frequency, which is defined as half the sampling rate. In the region adjacent to the maximum frequency, there is a possibility of phase distortion, so it is necessary to use a higher sampling rate with some margin for reliable measurements. In this study, the Jetson Nano alone could not support all the functions. As a secondary embedded system, an Arduino R1 was used to measure the Inertial Measurement Unit (IMU) located in the axle box via an I^2C interface.

Figure 9:　　　**Test Jig for Simple Modal Testing and Analysis**

If the Arduino R1 focuses on measuring vibration data from only one IMU, it can reliably support sampling rates between 1,100 and 1,200 Hz. Finally, a sampling rate of 1,024 Hz, a multiple of 2, was used, and it can reliably measure vibration data up to the Nyquist frequency of 512 Hz.

Although the importance of bending modes in wheelsets is known from the literature, it is necessary to verify whether the bending modes in 3D-printed wheelsets are present in the measurable frequency domain or not. If they are not, it would be the same as using a rigid wheelset. For numerical approaches using the Finite Element Method (FEM), this is not easy due to difficulties in directly modelling the infill structure patterns used in FDM 3D-printed parts and the lack of accurate material properties from filament manufacturers. Under the assumption that there are no major errors in the modelling process, the only information that can be obtained from numerical modal analysis using FEM simulation is the mode shapes and their order, as shown in Figure 8. The respective resonance frequencies and damping ratios are not accurate.

The alternative to insufficient information is to perform experimental modal testing and analysis, but there is no test jig, accelerometer, impact hammer, and software, etc. A simple testing jig was built using aluminium profiles to implement free-free conditions for the wheelset, as shown in Figure 9. The wheelset is suspended using a flexible hanger fabricated from TPU filament. An IMU, used for measuring vibration data in the axle box, were installed on the sides of the wheel, and the same position was structurally excited based on prior knowledge of the mode shapes. Due to the weight of the IMU relative to that of the wheelset and the lack of an impact hammer, this is not accurate, but high vibration responses at 200 Hz and 400 Hz are present and are believed to be the first and second bending modes. With this information, the finite element model of the wheelset was calibrated. Table 6 shows the calibrated material properties by part. Table 7 shows the resonance frequencies estimated from the numerical and experimental modal analysis. As a result, the effects of the two bending modes can be adequately measured within the maximum measurable frequency of 512 Hz.

Part Name	Young's Modulus	Poisson's Ratio	Mass	Volume	Density
Wheel	650 MPa	0.35	27.3 g	4.899e-5 m3	560.9 kg/m3
Axle	650 MPa	0.35	19.0 g	3.388e-5 m3	557.3 kg/m3

Table 6:   **Material Properties of PLA Filament by Part**

Mode Number	Natural Frequency	Mode Shape Description
1-6	-	Rigid Body Modes
7	162.28 Hz	First Torsion Mode
8	200.76 Hz	First Bending mode
9	398.68 Hz	Second Bending Mode
10	756.88 Hz	Third Bending Mode

Table 7:   **Modal Frequencies and Descriptions of Vibration Modes**

3.1.3   Wheel Flats with Different Modelling Methods

In the railway vehicle-track dynamics models, simplified 2D models of wheel flats have been mainly used for computational efficiency. However, these are not directly applicable to a 3D-printed wheel. Depending on its 3D geometry, it is expected to produce different excitation forces at the wheel-rail contact. To address this problem, three methods for modelling the 3D geometry are proposed, as illustrated in Figure 10: Type 1 – Simple Cut, Type 2 – Boolean Operation, and Type 3 – FEM Simulation. The size of the wheel flat was defined by its depth. The critical levels of 10 mm depth and 50 mm length are reduced to 1 mm and 5 mm respectively at 1/10 scale by the law of similarity. These values were initially considered, but due to the higher damping ratio of PLA compared to steel, a depth of 2 mm, which is twice as large, was commonly used to obtain distinct vibration responses in three modelling methods.

The first method 'Type 1 – Simple Cut' is to cut out the wheel flat area as if projecting a traditional 2D wheel flat model with sharp edges onto a 3D-printed wheel. Figure 10 (a) visually demonstrates this approach. In Autodesk Fusion 360, it is easy to change the depth to create a defective wheel with the desired depth of a single wheel flat. Once the CAD model is converted to STL format, it can be printed directly using slicer software for FDM 3D-printing. The printing time takes about two hours, depending on the performance of 3D-printers. The modelling is easy, but the created 3D geometry is closer to the dynamics model than the wheel of the real metro vehicle. Also, when cutting the wheel flat, both the wheel's tread and flange are trimmed, which seems to be a risk of derailment on a curved track.

The second method 'Type 2 – Boolean Operation' uses a Boolean operation (subtraction) rather than a simple cut to model the 3D geometry of the wheel flat more realistically and avoid trimming the wheel flange. Figure 10 (b) shows that the geometry of the wheel flats looks more realistic. The boundaries of the subtracted regions are sharp, but it is expected that they will become rounded as they wear away during repeated dynamic testing. The CAD modelling and printing process is the same as for Type 1 - Simple, and the time and effort required is similar. Unlike Type 1 - Simple Cut mentioned above, there seems to be no risk of derailment on a curved track. During dynamic testing, if the wheel and rail are not aligned as in the Boolean operation, a lateral excitation force is expected.

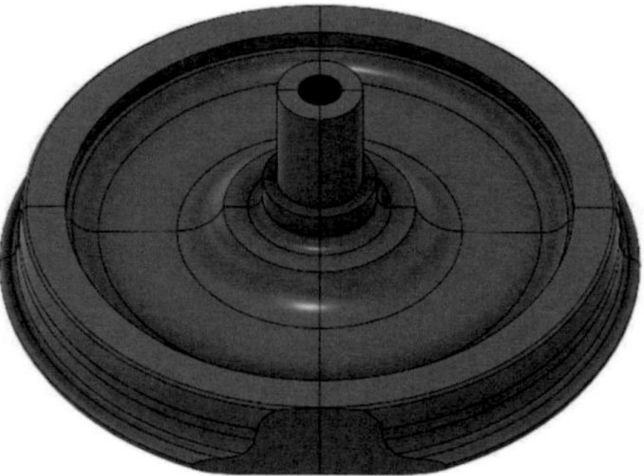

(a) Wheel Flat: Type 1 - Simple Cut

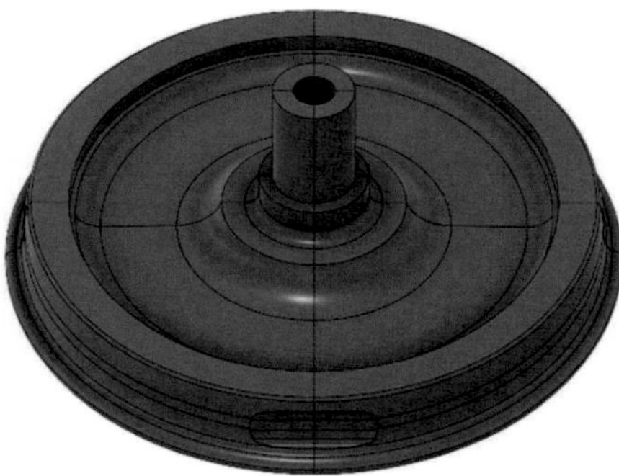

(b) Wheel Flat: Type 2 - Boolean Operation

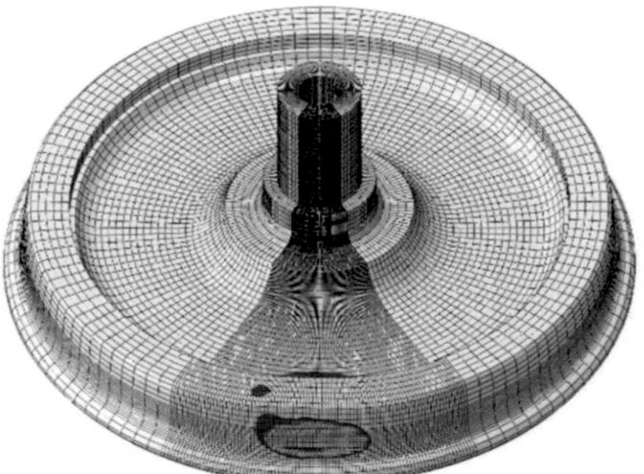

(c) Wheel Flat: Type 3 - FEM Simulation

Figure 10:   **Three Modelling Methods for Implementation of 3D Wheel Flats**

The third method 'Type 3 – FEM Simulation' creates a more realistic 3D geometry using FEM simulation than the previous two methods. The simulation aims to mimic the actual occurrence of a wheel flat by skidding the wheel on the rail while pressing down to the desired wheel flat depth. Plastic deformation on the wheel tread during skidding results in a wheel flat as shown in Figure 10 (c) and Figure 11.

The simulation was performed using Dassault Systèmes Abaqus. Firstly, the non-defective wheels and simplified rail were exported in STL format from Autodesk Fusion 360. After importing them in Abaqus, they were discretised using second order hexahedral elements (C3D20). The mesh consists of 383,130 nodes and 82,170 elements. For stable wheel-rail contact and computational efficiency, the mesh was set to be fine in the region of wheel angle +-30 degrees around the contact point of the wheel and rail. The friction between the wheel and rail is set to 0.3. Material properties were applied with Young's modulus, Poisson's ratio, density, and stress-strain curves as shown in Table 8 and Table 9. Although the wheel in the scale model is made of PLA filament, the material properties and stress-strain curves of common steel used in the tutorials in Abaqus are used since the aim is to simulate the plastic deformation in a real wheel. The simulation was performed in a quasi-static state with the wheel skidding on the rails. The calculation took about 5 hours using Intel i9-9900 CPU. The deformed wheel can be exported in STL format from Abaqus and printed in the same way as the previous two types. The 3D geometry of this realised wheel flat looks visually smaller compared to the previous two types.

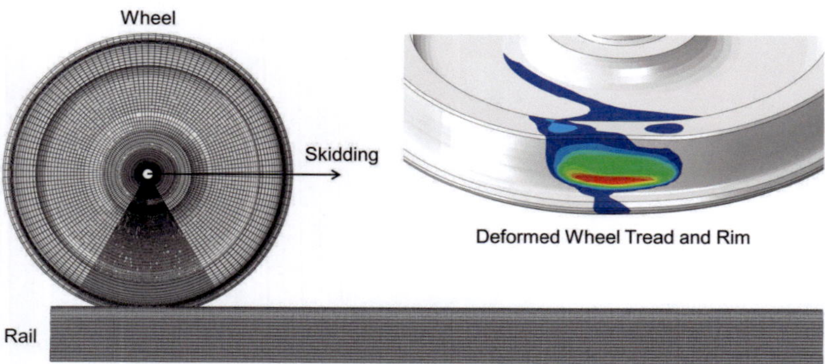

Figure 11:   **Modelling of Wheel Flat: Type 3 - FEM Simulation**

Parameter Name	Young's Modulus	Poisson's Ratio	Density
Value	210,000 MPa	0.3	7850 kg/m3

Table 8:   **Material Properties of Steel**

Plastic Strain	0.0000	0.0235	0.0474	0.0935	0.1377	0.1800
Yield Stress	200 MPa	240 MPa	280 MPa	340 MPa	380 MPa	400 MPa

Table 9:   **Stress-Strain Curve for Plastic Deformation of Wheel**

3.2   Fused Deposition Modelling (FDM) 3D-Printing

In this study, almost all the structural parts were fabricated using low-cost desktop Fused Deposition Modelling (FDM) 3D-printers (Appendix II: Fused Deposition Modelling 3D-Printers) and PLA filament, except for a few parts such as the bogie central frame, which required high stiffness and strength as the main skeleton. FDM 3D-printing is a popular additive manufacturing method that extrudes material layer by layer to build a three-dimensional object from the bottom up. The print head, which heats and extrudes the filament, moves along predefined paths to deposit the material precisely where it is needed. PLA is a biodegradable thermoplastic derived from renewable resources such as corn starch and is characterised by its ease of use, low cost, and good mechanical properties for non-industrial applications.

Initially, metal materials and machining were considered, but to reduce costs and time, especially for design changes, FDM 3D-printing was selected. This method allows for rapid, cost-effective prototype production in the lab. Desktop FDM printers are used despite their lower quality compared to industrial printers. The printing orientation was chosen to handle expected loads, and despite temperature-related print quality variations, a stable printing environment helped ensure consistency. For the parts' infill, a gyroid pattern with isotropic characteristics was chosen, balancing structural integrity and weight, despite the layering process's inherent anisotropy. The Gyroid pattern provides uniform strength and uses less material, making it cost-effective. It was initially intended to be produced with metal materials and through metal machining methods. However, this traditional manufacturing approach, involving outsourcing and high costs of materials and machining, presents significant limitations. Any design modifications due to flaws or the need to implement new design ideas quickly become burdensome in terms of time and cost. In contrast, FDM 3D-printing within the lab circumvents these issues, enabling swift and cost-effective iterations.

This decision was driven by the cost-effective and efficient nature of 3D-printing, which allows for rapid prototyping and an iterative design process in a laboratory environment. Low-cost desktop FDM 3D-printers were deemed sufficient for the requirements of this study, although the print quality is slightly lower and more scattered compared to industrial-level FDM 3D-printers. Figure 12 shows that the Gyroid infill structure is commonly used in the construction of the scale model. Figure 12 shows the bogie model on the track constructed by using FDM 3D-printing.

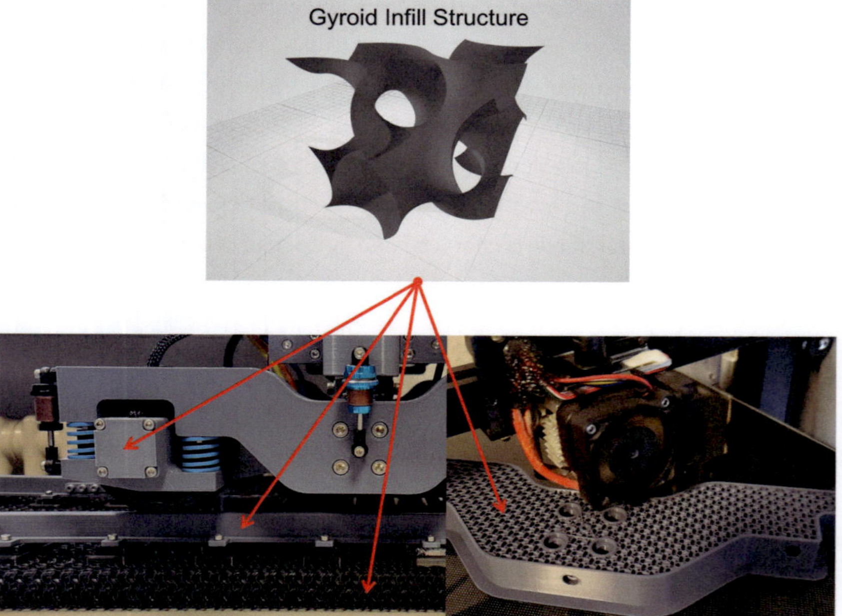

Figure 12:   Use of Gyroid Infill Structure in FDM 3D-Printed Parts

(a) Bogie Model on Track

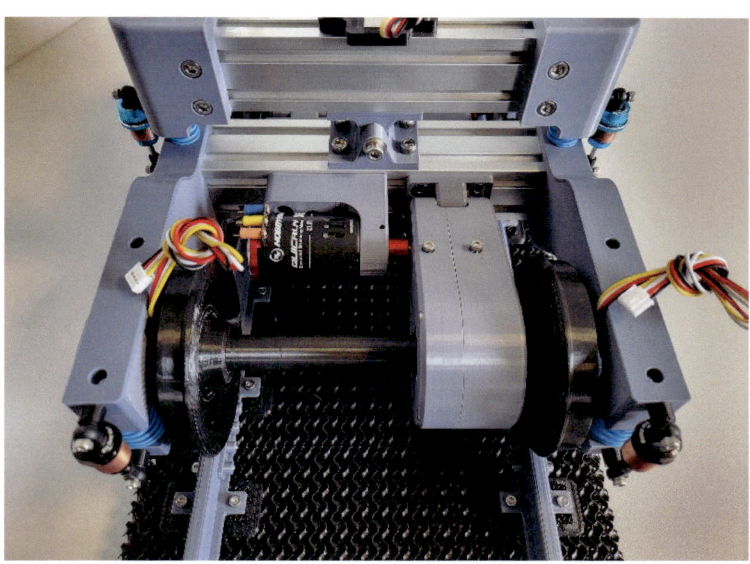

(b) Propulsion System

(c) Primary Suspension and Axle Box

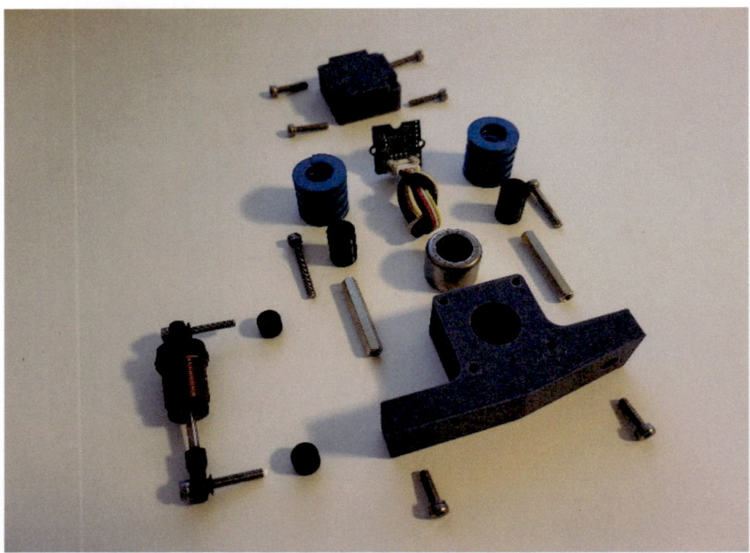

(d) Disassembled Axle Box

Figure 13:   **1/10 Scale Bogie Model on Track Constructed using FDM 3D-Printing**

3.3   Measurement and Control System

This system aims to localise the vehicle on the track, control its speed, and measure vibration data from the axle box with adjacent with wheel flats. Figure 14 shows the layout of the measurement and control system. In the centre of the control box, NIVIDIA Jetson Nano, one of the popular low-cost embedded systems, is installed and can be controlled by a client computer via a wireless network. The main functions are as follows:

The first function is to localise the vehicle on the track. An IMU, specifically LSM6DS3, installed in the centre of the control box is connected to Jetson Nano via I^2C interface, and information about the vehicle's movements can be obtained by integrating 3-axis accelerations. On the track, magnets are installed at intervals of 297.5 mm, the length of the track module, and a hall effect sensor installed at the bottom of the bogie frame is connected to the digital input port, and its movement on the track can be estimated from the digital pulses obtained when the vehicle passes over the magnets. The hall effect sensor embedded in the BLDC motor is also connected to the digital input port and can estimate the vehicle's movement based on the motor speed.

The second function is to control the speed of the vehicle. Based on the information about the vehicle position on the track, the vehicle speed can be controlled by adjusting Pulse Width Modulation (PWM) to the controllers of the two BLDC motors installed before and after the bogie.

The third function is to measure vibration data from the axle box. During the iterative process of determining the vehicle position and controlling the motors, it is difficult for Jetson Nano alone to reliably measure vibration data from the IMU installed in the axle box at a sampling rate of 1024 Hz, which may be too high for a low-cost embedded system. As an alternative, an Arduino R1 was used to assist the Jetson Nano. Arduino R1 and Jetson Nano are connected to each other via serial communication using a USB cable. Figure 15 illustrates the process flow for measuring vibration data from the axle box using the combination of Arduino R1 and Jetson Nano. The measurement begins when a request is sent from the Jetson Nano to the Arduino via serial communication, prompting the initiation of the data capture sequence.

Firstly, the Arduino prepares for data collection by initializing a measurement buffer within its internal memory. This buffer is a dedicated space where the incoming vibration data, in the form of 3-axis accelerations detected by the IMU, will be temporarily stored. Upon successful initialisation, the Arduino proceeds to continuously read the 3-axis acceleration data from the IMU. This step is crucial as it captures the raw vibrational signals that are necessary for analysis. These readings are then saved to the previously initialised measurement buffer. This iterative process is sustained, ensuring that all relevant data are captured and stored sequentially. To ensure data integrity and consistency, the Arduino adjusts the timing of the measurement loop. This adjustment is vital to stabilise the sampling rate, which is essential for maintaining the fidelity of the time series data and for allowing accurate post-processing and analysis. The system then checks if the measurement cycle is complete. If it is not, the loop continues, reading and storing subsequent data.

If the measurement is complete, the process progresses to the next step. Once the data collection is deemed complete, the Arduino sends the vibration data back to the Jetson Nano via serial communication. The Jetson Nano then takes over to handle the data, presumably for further processing or real-time analysis. The final step involves writing the collected vibration data to a CSV file. This format is chosen for its simplicity and compatibility with various data analysis tools and environments. The CSV file serves as a permanent record of the measurements, enabling detailed analysis and archiving for future reference. The process concludes with the measurement being marked as finished, signifying that the data is now ready for review and the system is ready for the next measurement cycle if required. This structured and systematic approach ensures the accurate capture of high-frequency vibration data from the axle box, which is critical for diagnosing issues like wheel flats and assessing the overall health of the railway vehicle's wheel-rail interface.

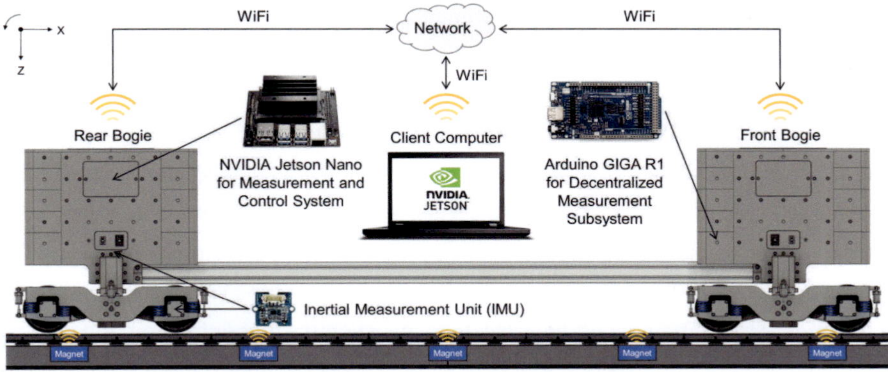

Figure 14: **Overview of Measurement and Control System**

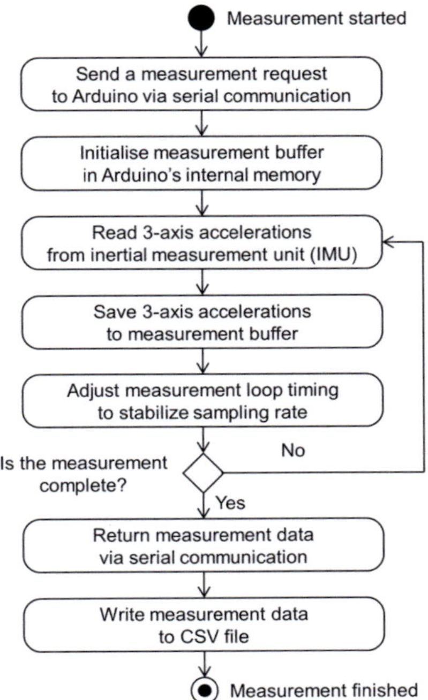

Figure 15: **Measurement Process of Vibration Data from Axle Box**

3.4   Dynamic Testing and Vibration Analysis on Short Straight Track

Dynamic testing was carried out to experimentally measure vibration data from the axle box at the level of the bogie model, not at the level of the vehicle with two bogie models, on a short straight track of 2.4 metres in length. Figure 16 shows the detailed design and 3D-printed bogie model with a single wheel flat in the front view. This experiment has two purposes. The first purpose is to prepare input datasets to identify the occurrence and types of wheel flats using a classification model, and measurements were performed for a total of four conditions: Non-defect, Type 1 - Simple Cut, Type 2 - Boolean Operation, Type 3 - FEM simulation. The depth of the wheel flats was set to a common 2 mm. The second objective is to investigate whether the first and second bending modes of the wheelset, estimated to be located at approximately 200 Hz and 400 Hz, significantly affect vibration data from the axle box.

The Bogie model is controlled by Jetson Nano to run forward and backward round trips on the track with a speed profile of acceleration, constant speed, and deceleration. At constant speed, it runs at up to 1 m/s. The position on the track is estimated by integrating the x, y, and z accelerations measured by the IMU and sensing the magnets placed at regular intervals of 295 mm on the track with Hall effect sensors placed under the bogie frame. Based on this information, the speed of the bogie model is controlled, and a piezo buzzer is used to provide auditory feedback on its operation. Arduino R1 was then used to measure vibration data from an IMU installed in the axle box adjacent to the wheel with wheel flats. Jetson Nano is not enough to cover all the functions, especially for high sampling rates. Due to the limited processor speed, Arduino R1 can provide a sampling rate of up to about 1,100-1,200 Hz over the I^2C interface. The sampling rate was chosen to be 1,024 Hz, a power of two according to Nyquist's theory, with a slight frequency margin to allow the vibration data to include the second bending mode, which is estimated to be located at around 400 Hz.

Jetson Nano and Arduino R1 are connected to each other via serial communication using a USB cable. Before the bogie model starts running at one end of the track, Jetson Nano sends the desired measurement time to Arduino R1 via serial communication, where the desired time is set to 20 seconds. Then the bogie model starts to run. During the test, the vibration data from the IMU is stored in the Arduino R1 internal memory. After the test is over, the four columns of data - time, x-, y-, and z-axis accelerations - are transferred to Jetson Nano at once via serial communication and saved as a CSV format file. Arduino R1 has a Real Time Clock (RTC), but it is not preserved after power off. Therefore, when a measurement is started, a reference time is obtained over the internet from the Network Time Protocol (NTP) server. Since the time resolution is 1 second, the rest of the time was recorded by the RTC with a time step of 1/1024 second. The dynamic testing was performed with 25 repetitions per condition. Postprocessing was performed to analyse and compare them. The vibration data consists of x-, y-, and z-axis accelerations of 20 seconds in length.

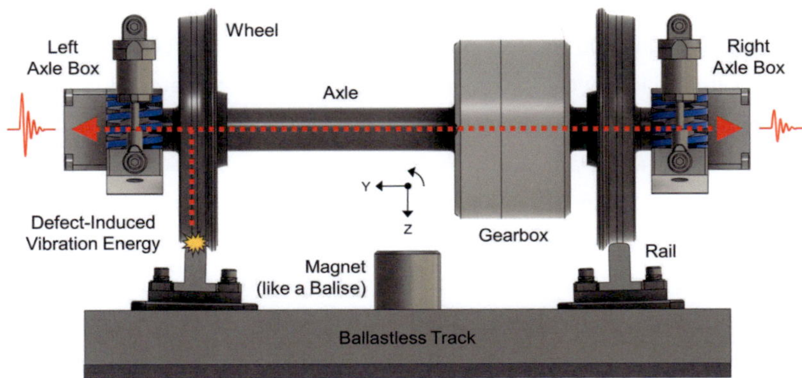

(a) Detailed Design

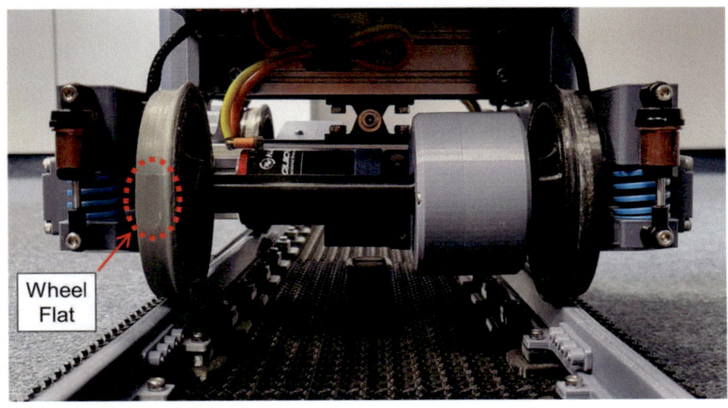

(b) 3D-Printed Bogie Model

Figure 16:   **Detailed Design and 3D-Printed Bogie Model with Wheel Flat**

Figure 17 shows a time series of 5 seconds in length per label. It also shows a time-frequency map calculated using the CWT. The Fourier transform, which uses the sine and cosine functions as shape functions, is relatively difficult to capture impulsive components in vibration data like wavelet functions. Therefore, CWT with the Morlet wavelet function was applied to the vibration signal. If the Short-time Fourier Transform (STFT) was applied, the time-frequency map would show a tendency for the results to be clumpy. This calculation was performed with 'pywt' library in Python. Figure 18 shows the mean and 95% confidence intervals for the Power Spectral density (PSD) of the 25 vibration signals per label.

The measurement results commonly show the existence of a large vibration response for all labels at around 200 Hz. This is presumably due to the first bending mode of the wheelset, with little influence from the second bending mode at around 200 Hz. Even in case of non-defect, a large vibration response at 200 Hz is clearly visible. In addition to wheel flats, track irregularities also excite wheels and the bogie model. In particular, the rails were printed with 295 mm in length due to the limited printable size of FDM 3D-printers. It is presumed that the discontinuities at the rail connections were not smooth and were constantly exerting forces on the wheelset. These track irregularities are expected to cause the wheelset to have a first bending mode and the bogie model to have a bouncing mode in terms of vehicle dynamics. Several resonances in the frequency domain below 100 Hz were assumed to be due to vehicle dynamics, flexible tracks, and rigid bodies, but are actually due to flexible vehicle structures and need to be further characterised in future studies.

Type 1 - Simple Cut shows a clear and consistent impulse with each rotation of the wheel, but the geometry of the wheel flat is not close to reality. Type 2 - Boolean Operation and Type 3 - FEM Simulation show less consistent impulses at each rotation compared to Type 1 - Simple Cut, but the geometry of the wheel flats is closer to reality. Which modelling method is suitable for follow-up studies will depend on the criteria considered important, but at least these results show that different modelling methods of wheel flats result in different vibration signals.

In Chapter 5, after investigating how different response characteristics are classified by the classification models, there will be a discussion of which modelling methods are appropriate for follow-up studies.

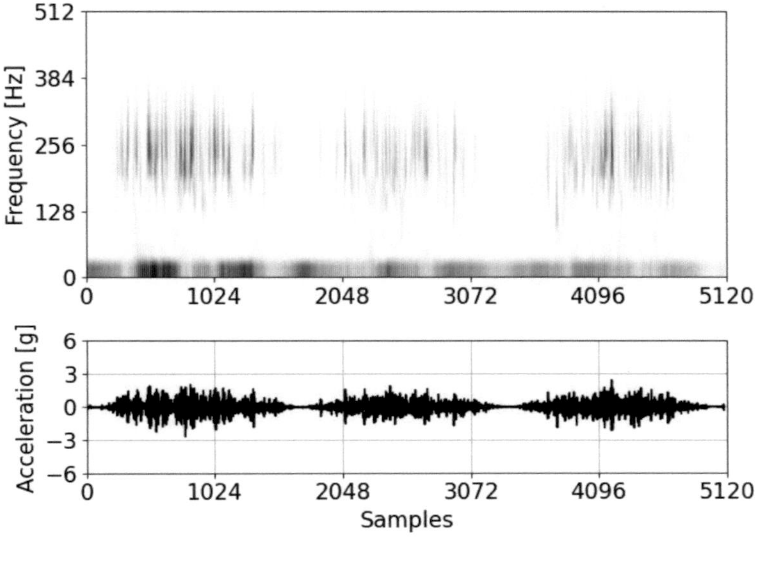

(a) Non-defect

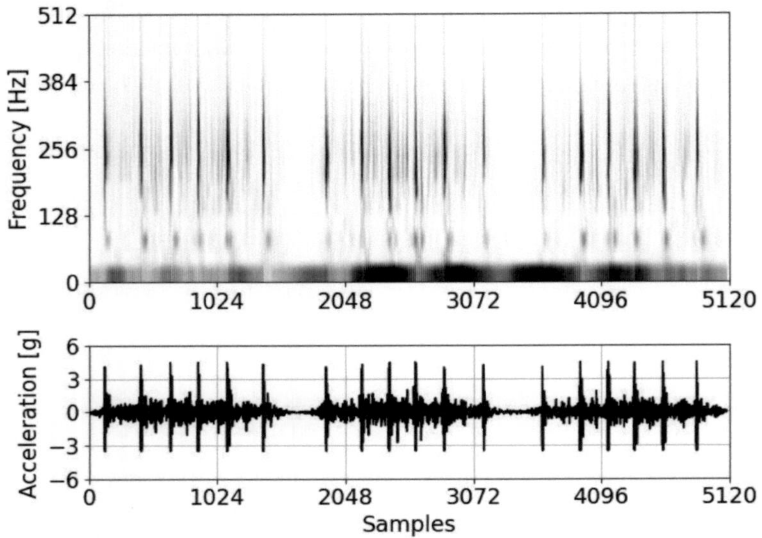

(b) Wheel Flat: Type 1 - Simple Cut

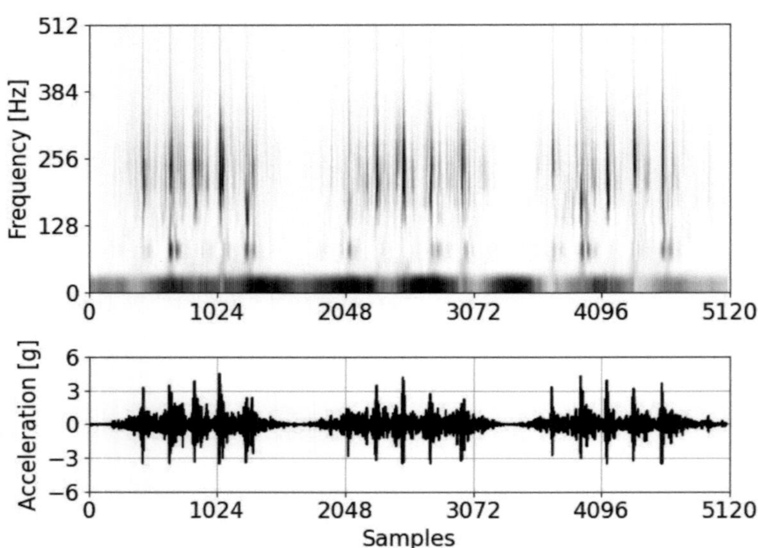

(c) Wheel Flat: Type 2 - Boolean Operation

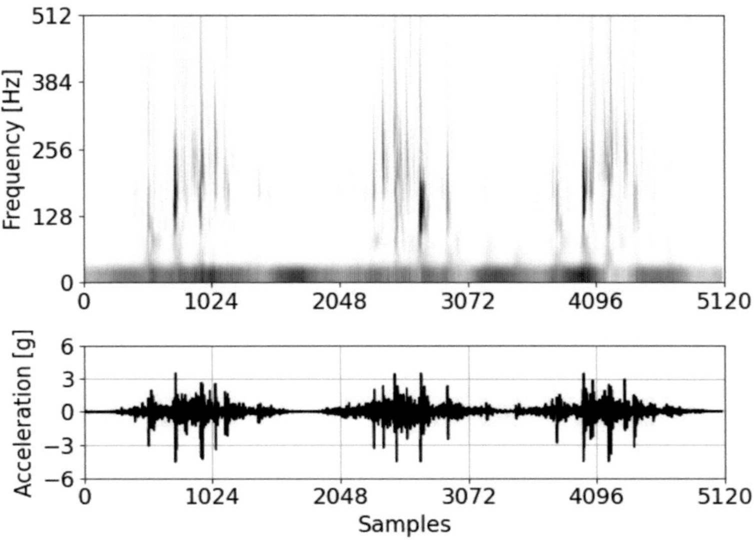

(d) Wheel Flat: Type 3 - FEM Simulation

Figure 17: **Time-series Signal and Continuous Wavelet Transform (CWT) by Wheel Flat Geometry**

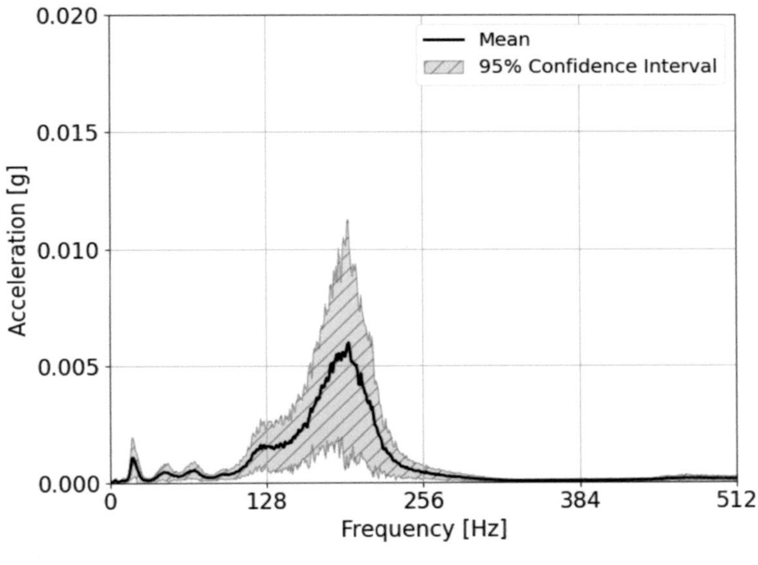

(a) Non-defect

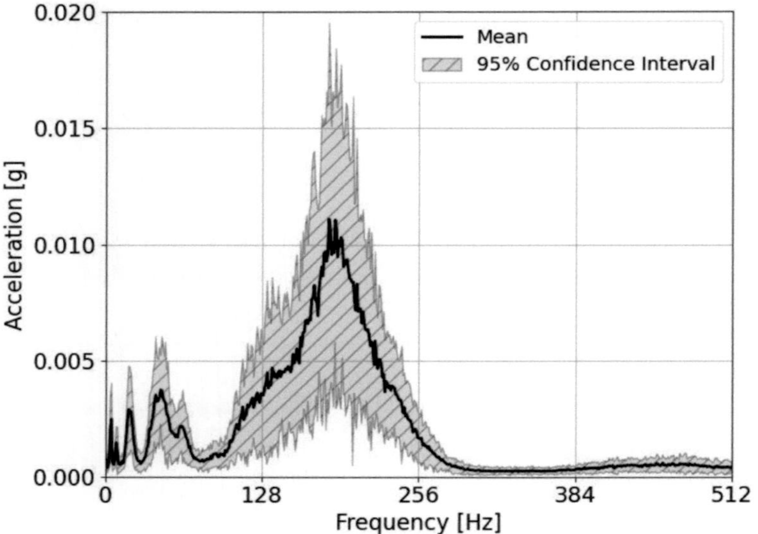

(b) Wheel Flat: Type 1 - Simple Cut

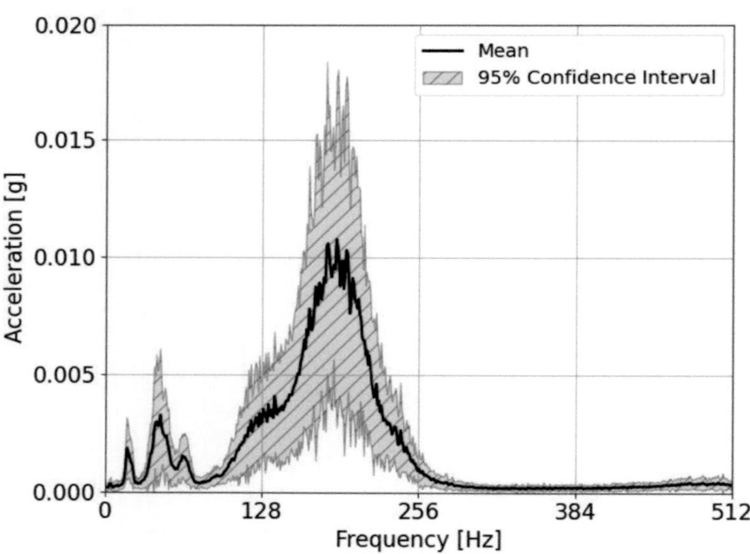

(c) Wheel Flat: Type 2 - Boolean Operation

3D-Printed Scale Model for Detection of Railway Wheel Flats using Augmented Vibration Data from Axle Box

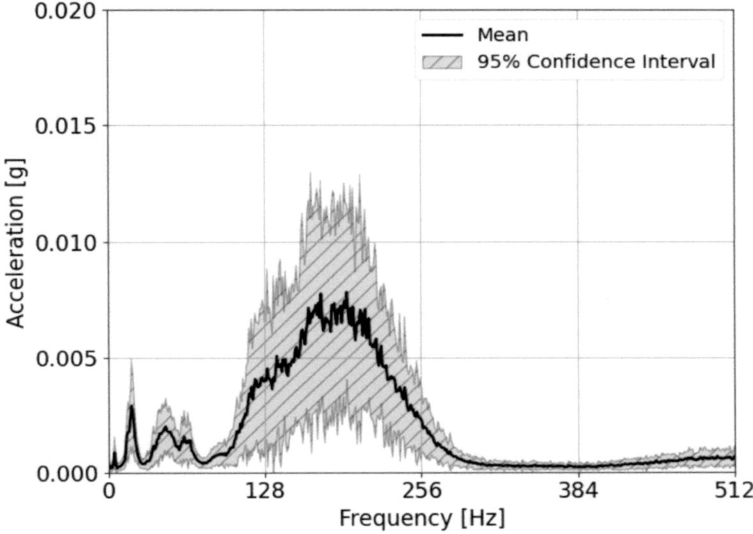

(d) Wheel Flat: Type 3 - FEM Simulation

Figure 18: **Power Spectral Density and 95% Confidence Intervals by Wheel Flat Geometry**

4 Generative Model-based Vibration Data Augmentation for Vertical Axle Box Accelerations

This chapter introduces how to preprocess and augment the raw vibration data obtained experimentally from the 3D-printed railway scale model. To generate synthetic vibration data using a generative model based on preprocessed vibration data, the following tasks were performed: 1) preprocessing the raw vibration data with signal segmentation and high-pass filtering, 2) modelling the generative model using a LSTM network, 3) optimising the hyperparameters of the generative model by employing a genetic algorithm, 4) augmenting and validating the synthetic vibration data based on the preprocessed vibration data, and 5) discussing the results.

4.1 Preprocessing of Raw Vibration Data

The raw vibration data measured by an IMU on the axle box, adjacent to a wheel flat, was preprocessed so that a generative model could better learn and distinguish between normal and abnormal vibration patterns caused by the three different geometries of the wheel flats. It includes 3-axis accelerations in the x, y, z directions. The preprocessing was only applied to the vertical vibration data because the excitation forces induced by the wheel flats at the wheel-rail contact were assumed to be dominant in the vertical direction, and in fact their vibration energy is dominates the other two directions, lateral and longitudinal, in the measured vibration data. As a result, a total of 800 segments for four labels in the vertical direction, 200 segments per label, were used as the input dataset to train, validate, and test a generative model.

Figure 19 shows that the detailed preprocessing starts with reading the 25 measurement files per label in the CSV format. To remove the effects of vehicle dynamics and the tilt of a capacitor-type IMU sensor, which affect the 3-axis accelerations at low frequencies, the 5th-order Butterworth high-pass filter was applied as it has a sharp cut-off and minimal ripple in the pass band to avoid distorting the raw vibration data. The cut-off frequency was set at 50 Hz based on the low frequency energy. From here on, the filtered accelerations in the vertical direction will be referred to as the filtered vibration signals without specifying the direction. The filtering process was carried out using the 'butter' and 'filtfilt' functions from the 'scipy.signal' module in Python.

The top graph in Figure 20 shows the filtered vibration signal which was measured at the axle box as the bogie model repeatedly runs forward and backward on the short straight track of 2.4 m in length. After high-pass filtering, the DC component of the filtered vibration signal is at zero, and there are no more low-frequency oscillations due to vehicle dynamics. However, due to acceleration, deceleration, and changes of driving direction, the dominant defect-induced energy in the filtered vibration signals is inconsistent and vary with operating conditions. Therefore, instead of using them directly to train, validate, and test a generative model, they were segmented centred on the indexes of impulses induced by the wheel flats and then filtered to remove the segments with low energy magnitude and low Signal-to-Noise Ratio (SNR). The indexes of impulses within the filtered vibration signals were searched using the 'argmax' function from the 'NumPy' module in Python. The length of the segment was determined to be 256 samples to account for the distance between impulses. Since each signal contained 40 impulses, 40 segments per signal were extracted. In Figure 20, the bottom left graph shows the overlap of the 1,000 segments for label 2.

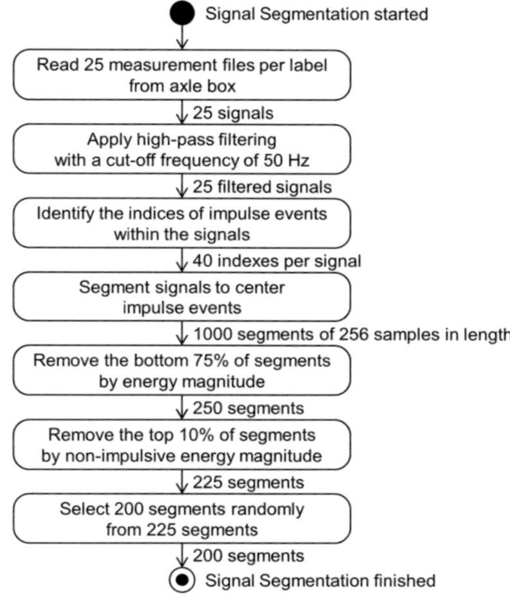

Figure 19:   **Signal Segmentation and Filtering Process**

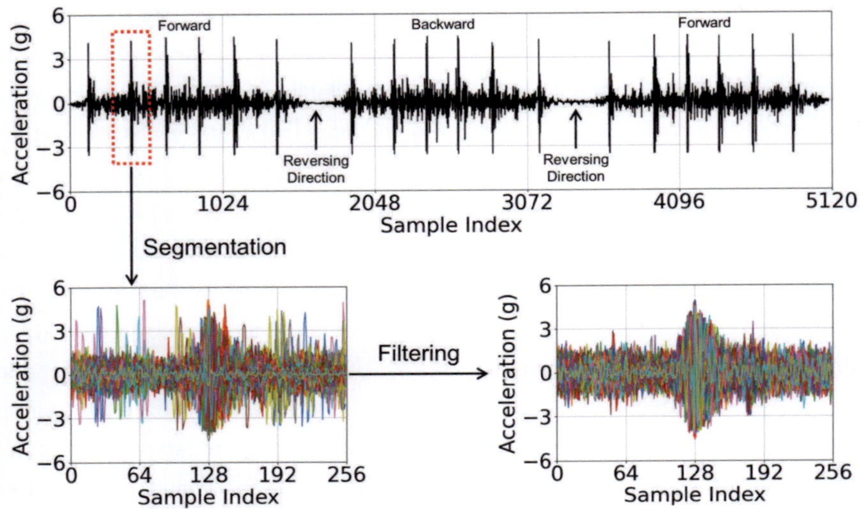

Figure 20:   **Example of Signal Segmentation and Filtering**

To make the segments more consistent and have a higher SNR, the lower 75% of the segments with low vibration energy were first removed, where the energy of each segment is calculated using the sum of the squares of its sample values according to the equation:

$$E(x) = \sum_{i=1}^{N} x_i^2 \tag{1}$$

where $E(x)$ is the energy of the segment and x_i represents the amplitude of the segment at the ith sample. This equation provides a scalar measure of the energy content within a segment, which is used to sort the segments in terms of their signal energy magnitude. These are segments with relatively low vibration energy, mainly during acceleration, deceleration, and changes in driving direction. Then, the top 10% of segments with high vibration energy in the side regions without the central 40 samples, defined as non-impulsive vibration energy, were removed. This filtering process filtered out 775 of the 1,000 segments per label. The filtering criteria of lower 75% and top 10% were chosen by trial and error to achieve the desired results. In Figure 20, the bottom right graph shows the overlap of the 225 filtered segments.

To enhance the generalisation capability of a generative model and to avoide over-fitting, 200 out of the 225 segments per label were randomly selected to provide a slightly varied input dataset for a generative model. This approach not only exposes a generative model to a wider range of data patterns, thereby strengthening its capability to generalise, but also helps to reduce the risk of a generative model becoming over-fitted to a particular subset of the vibration data.

4.2   Modelling of Generative Model using Long Short-Term Memory Network

4.2.1   Long Short-Term Memory (LSTM) Network

An LSTM network is used as a generative model for data augmentation of the prepro-cessed vibration signals. Figure 21 shows the architecture of a simple LSTM network consisting of a cell and a dense layer, where a cell is designed to extract high-level features from time-series data where temporal sequences are important in the input data, and its output is mapped through a dense layer to the final low-level output in the desired format. A cell serves as the core structural unit which is governed by gates in an LSTM network: the forget gate f_t, the input gate i_t , and the output gate o_t. These gates are regulated by weights and biases, which are adjusted during the training pro-cess.

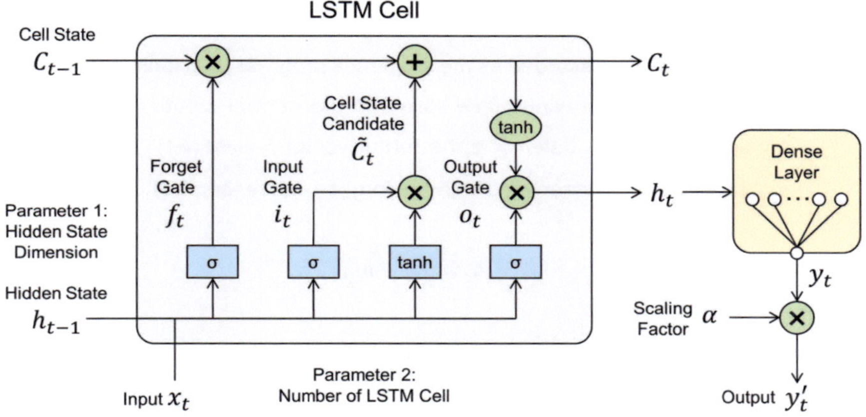

Figure 21:   **Architecture of LSTM Network with Energy Correction Factor**

They can selectively allow information to flow through, be modified, or be forgotten across time steps. This mechanism effectively overcomes the vanishing gradient problem commonly encountered in traditional recurrent neural networks (RNNs). As a result, LSTM networks can learn and retain information over longer sequences, making them particularly suitable for tasks that involve extended temporal dependencies. They have been widely used for various purposes, ranging from time-series forecasting, natural language processing, and many other applications. The mathematical operations of a cell, a dense layer, and a scaling factor can be described as follows:

- **Input and Output Layers:** Both layers were configured with a single-channel input x_t and output y_t to accommodate the dimensions of the segments extracted from the preprocessed vibration data. Within the LSTM cell, the dimensions of input and output are intricately linked to the dimensions of the hidden state h_{t-1}. This hidden state h_{t-1} plays a pivotal role in the model, as it significantly influences its capability to capture complex patterns in the preprocessed vibration data. Choosing an appropriate dimension for the hidden state h_{t-1} is a critical decision, as it balances the model's learning capacity and computational efficiency, while protecting against overfitting. Consequently, the dimensions of the hidden state h_{t-1} are crucial hyperparameters within the cell.

- **Forget Gate:** The forget gate f_t decides which parts of the cell state C_{t-1} should be remembered or discarded as the sequence progresses. It looks at the hidden state h_{t-1} from the previous time step and the current input x_t and applies a sigmoid function σ to determine the forget vector f_t, where W_f and b_f are the weights and biases associated with the forget gate, respectively.

$$f_t = \sigma\big(W_f \cdot [h_{t-1}, x_t] + b_f\big) \tag{2}$$

- **Input Gate:** The input gate i_t and a set of candidate values $\tilde{C}_t$ together decide what new information will be stored in the cell state. It consists of two parts: a sigmoid layer σ that creates the input gate, and a hyperbolic tangent layer $tanh$

that creates a vector of candidate values $\tilde{C}_t$, where W_i, b_i, W_C, and W_C are the weights and biases for the input gate and the creation of candidate values, respectively.

$$i_t = \sigma(W_i \cdot [h_{t-1}, x_t] + b_i) \tag{3}$$

$$\tilde{C}_t = tanh(W_C \cdot [h_{t-1}, x_t] + b_C) \tag{4}$$

- **Cell State Update:** The cell state C_t is the memory part of the LSTM cell. The old state C_{t-1} is updated to the new cell state C_t by combining a set of candidate values $\tilde{C}_t$. This update is a combination of forgetting the old values signalled by the forget gate and adding the new candidate values signalled by the input gate.

$$C_t = f_t * C_{t-1} + i_t * \tilde{C}_t \tag{5}$$

- **Output Gate:** The output gate o_t decides what the next hidden state h_t should be. The cell state C_t is passed through a hyperbolic tangent function $tanh$ to ensure the output values stay between -1 and 1, and then multiplied by the output of the sigmoid gate function σ to decide which parts of the cell state will output, where W_o and b_o are the weights and biases for the output gate.

$$o_t = \sigma(W_o \cdot [h_{t-1}, x_t] + b_o) \tag{6}$$

$$h_t = o_t * tanh(C_t) \tag{7}$$

- **Dense Layer:** This layer is fully connected to the output of the LSTM cell and is responsible for transforming the high-level features automatically extracted from training within the LSTM cell into the desired output form. The Rectified Linear Unit (ReLU) activation function is employed here to introduce non-linearity, which is essential for the network's ability to learn complex patterns. The ReLU outputs the input directly if it is positive, otherwise zero. The output y_t is calculated by applying the ReLU activation function to the linear transformation of the hidden state h_t with weights W_d and biases b_d.

$$y_t = ReLU(W_d \cdot h_t + b_d) \tag{8}$$

- **Energy Correction Factor:** The energy correction factor α was applied externally to the output of the dense layer to match the energy magnitude of the synthetic vibration data with that of the preprocessed vibration data. This addresses the problem where the synthetic vibration data is qualitatively similar to the input vibration signals, but does not match in terms of average energy magnitude. In the initial phase, a separate learning rate and the Adam optimiser were initially employed with the LSTM cell during training. However, it was observed that the training rate had a significant impact on the results, and for the sake of efficiency without an iterative process, the energy correction factor α was applied externally to the LSTM cell. The energy correction factor α was calculated by dividing the average energy magnitude of the synthetic vibration data by that of the preprocessed vibration data. The calibrated output y'_t is obtained by multiplying the output y_t with the energy correction factor α.

$$\alpha = E_{preprocessed}(x_t)/E_{synthetic}(y_t) \tag{9}$$

$$y'_t = \alpha \cdot y_t \tag{10}$$

- **Hyperparameter:** In configuring the hyperparameters of the LSTM cell, two hyperparameters were chosen that are important to enhance the quality of the synthetic vibration data: the size of the hidden state vector, which determines the width of the LSTM network, and the number of stacked LSTM cells, which determines the depth of the LSTM network. The size of the hidden state vector directly affects the amount of information the model can grasp and remember as the sequence progresses. A larger vector size helps the model learn more detailed patterns, which is important for handling the complexity of the vibration data and distinguishing between different labels. The number of stacked LSTM cells is crucial to the model's ability to understand complex dependencies within the sequential vibration data. This multi-layered approach allows the LSTM network to explore deeper into the temporal details, enabling it to distinguish small variations in the vibration data.

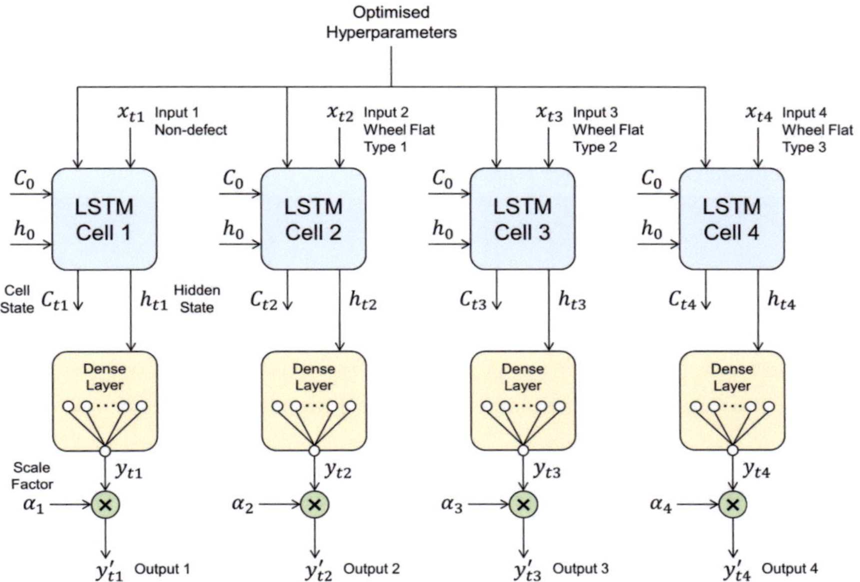

Figure 22: **Architecture of Parallelised LSTM Network**

4.2.2   Parallelised LSTM network and Implementation

The architecture of a parallelised LSTM was proposed to train, validate, and test multiple LSTM cells for four labels: non-defect and three types of wheel flats. Figure 22 shows the parallelised LSTM network. Each LSTM cell is trained independently and outputs a hidden state h_t, which is subsequently fed into a dense layer and scaled for energy magnitude matching. The parallelised LSTM network was implemented based on the LSTM network provided by the 'nn.Module' in the PyTorch framework, and then was trained using a NVIDIA RTX 2060 Super on a Window 10 system, utilizing Python 3.9 and PyTorch 2.1.

The training process beings by initialising the hidden h_0 and cell state c_0 to zero, for starting without any predefined weights and biases. Subsequently, the training loop then employs the Mean Squared Error (MSE) loss function, which calculates the mean squared difference between the actual and predicted values, serving as a critical metric for model accuracy. The Adam optimiser is utilised for its efficiency in adjusting network weights based on the computed gradients of the loss function, optimizing the training process. To mitigate overfitting and enhance model generalisation to unseen data, an early termination feature is integrated. This termination mechanism stops the training if there is no improvement in the validation loss over a predefined number of epochs, preserving the model's predictive reliability.

4.3   Hyperparameter Optimisation using Genetic Algorithm

It has been observed that the performance of the LSTM-based generative models is highly sensitive to the selection of hyperparameters. Hyperparameter optimisation is crucial in achieving a high correlation between the preprocessed and synthetic vibration data, as well as minimising the computational time required for training. In this study, a Genetic Algorithm (GA) is employed to optimise the hyperparameters due to its ability to avoid local optima and efficiently search for global optima, ensuring the discovery of the best possible solutions. The entire computational process was implemented in Python, without relying related open-source modules. Figure 23 shows the hyperparameter optimisation process using GA. The detailed explanation for each step is as follows:

The GA-based optimisation starts by defining hyperparameters and their ranges as shown in Table 10. A total of four hyperparameters were selected, considering the influence of each hyperparameter on the training process: the size of the hidden state vector, the number of stacked LSTM cells, the batch size, and the learning rate. Each range was determined by trial and error adjustments.

The initial population was randomly selected to maximise the diversity of genetic information of individuals within the predefined region of the hyperparameter space. The number of individuals per population was chosen to be 100. When an insufficient number of individuals was not used, convergence in the wrong direction was repeatedly observed. This is presumably due to the lack of individuals in the initial population and the bias of the genetic information to be evenly distributed within the range of the hyperparameters. However, considering more individuals is not easy due to the proportional increase in computation load. After trial and error, the number of individuals was determined to be 100 as a good balance point between computational load and convergent results.

As the first generation, the parallelised LSTM network is trained for all labels simultaneously over the initial population. The average training loss L_{avg} is calculated from the training losses L_{cell1}, L_{cell2}, L_{cell3}, L_{cell4} of the parallelised LSTM network as below:

$$L_{avg} = \frac{(L_{cell1} + L_{cell2} + L_{cell3} + L_{cell4})}{4} \tag{11}$$

The fitness function is defined in reciprocal form using the average training loss L_{avg} as below:

$$fitness = \frac{1}{1 + L_{avg}} \tag{12}$$

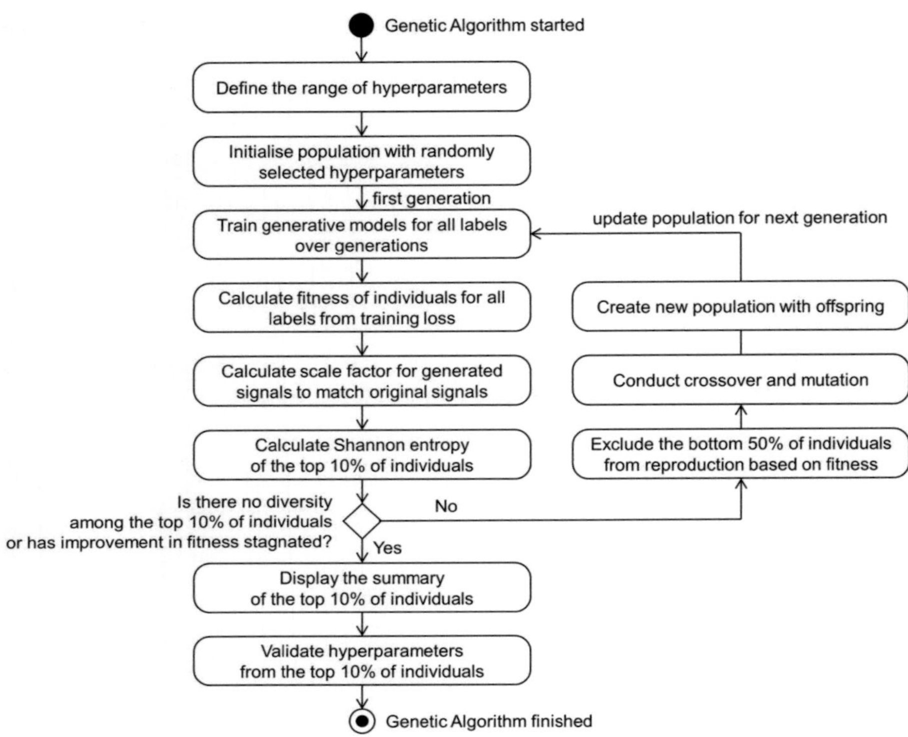

Figure 23: **Hyperparameter Optimisation Process for LSTM-based Generative Model using Genetic Algorithm**

The key features of this fitness function are its bounded nature, which keeps fitness within the range of 0 to 1, thus allowing clear comparisons between different individuals, and its non-linear scaling, which creates a clear distinction between better and worse individuals, significantly improving the selection process in the GA-based optimisation. Once the training and fitness evaluation of the initial population is complete, individuals are sorted within the population in descending order of fitness.

After training the LSTM network, 200 synthetic vibration segments per label are generated using random input data. They are compared with the same number of preprocessed vibration segments to evaluate their data quality. The scaling factor α is calculated by comparing both average energy magnitudes using Equation (9). This is not used directly to calculate the fitness, but be used as a secondary indicator to compare good and bad of trained LSTM networks.

The decision to continue or stop the evolutionary process is determined by the Shannon entropy, which is used to quantitatively measure the diversity of genetic information by hyperparameter within each individual as follows:

$$H(X_i) = -\sum_{j=1}^{n} P(x_{ij}) \cdot log_2\left(P(x_{ij})\right)$$

(13)

where $H(X_i)$ is the Shannon entropy for the i-th hyperparameter , X_i is the range of possible values that the i-th hyperparameter can take, n is the total number of district values that the i-th hyperparameter can take, i is an index indicating the hyperparameter number, j is an index running over the possible values within the range X_i, x_{ij} is a specific value within the range of possible values for the i-th hyperparameter $P(x_{ij})$ is the probability of the i-th hyperparameter taking the value x_{ij}. The total Shannon entropy $H_{total}(X)$ is the sum of the Shannon entropies $H(X_1)$, $H(X_2)$, $H(X_3)$, $H(X_4)$ for all hyperparameters.

$$H_{total}(X) = H(X_1) + H(X_2) + H(X_3) + H(X_4)$$

(14)

When there is no more diversity in the genetic information within the top 10% of individuals, the total Shannon entropy $H_{total}(X)$ becomes zero, meaning that the evolutionary process ends when the genetic information of individuals is the same. Then, the converged genetic information and evaluation results of the top 10% of individuals are displayed. Lastly, the converged genetic information was validated to see if it can generate the high quality of synthetic vibration data.

Otherwise, the evolutionary process will continue with a new population with new offspring through the crossover and mutation process based on only the top 50% of the population. As shown in Table 11, the crossover and mutation were performed with rates of 50% and 5%, respectively. The mutation rate was set at 5% as a means of exploring the potential for better genetic information through mutations before converging on specific genetic information with the Shannon entropy of 0. This rate was determined through trial and error to allow for the possibility of advantageous mutations that can enhance diversity and prevent premature convergence. In Table 11, the 'Fitness

Stagnation Threshold' and 'Maximum Stagnation Generations' are set values for early termination of the evolutionary process in situations where fitness stagnates.

Parameter Name	Parameter 1 Hidden State Size	Parameter 2 Number of LSTM Cells	Parameter 3 Batch Size	Parameter 4 Learning Rate
Range	1 2 4 8 16 32 64	1 2 4 8 16	8 16 32 64 128	1 0.1 0.01 0.001 0.0001

Table 10:   **Region of Hyperparameters for LSTM-based Generative Model**

Parameter Name	Crossover Rate	Mutation Rate	Fitness Stagnation Threshold	Maximum Stagnation Generations
Value	0.5	0.05	1e-3	10

Table 11:   **Hyperparameters for Genetic Algorithm**

4.4 Data Augmentation and Validation

To artificially enhance the dataset in terms of volume, variety, and velocity, based on the limited number of the preprocessed signals from the experiments, the data augmentation was performed using the LSTM-based generative model with the best hyperparameters obtained in the previous subchapter, following the process shown in Figure 24.

The first step is to read the best hyperparameters of the top 10% of individuals sorted by fitness in the evolutionary optimisation process using GA. The evolutionary process terminates when the genetic information shows no diversity among the top 10% of individuals. As a result, all individuals have the same hyperparameters. The next step is to set the desired number of synthetic vibration signals per label for a specific label.

When classifying the presence or absence of wheel flats and identifying their types, it is essential to determine the minimum number of synthetic vibration signals necessary to enhance the learning process and the classification performance. The number of synthetic vibration segments per label was configured in intervals of 200, ranging from 200 to 2,000.

Four LSTM cells and dense layers in the parallelised LSTM generative model were then trained separately for all labels with the best hyperparameters. After training, a random number generator was used to generate a set of randomised input data which equal in length to the segments and equal in number to the desired number of synthetic vibration data. The synthetic vibration data are then produced using the parallelised LSTM generative model for all labels. To ensure consistency in the energy magnitude $E(x)$ between the preprocessed vibration data and the synthetic vibration data, a scaling factor α is used to match the average energy magnitude $E(y')$ of the synthetic vibration segment $y'(t)$ to the average energy magnitude $E(x)$ of the preprocessed vibration segment $x(t)$ using Equation (1), (9), and (10). After being corrected using a scaling factor, there should be no more than a 1% difference ΔE in energy magnitude between both vibration data.

$$\Delta E = |E(x) - E(y')| \tag{15}$$

$$\Delta E \leq 0.01 \cdot E(x) \tag{16}$$

If not, the generative model starts learning again. If yes, the similarity between both vibration data is assessed using the following two metrics.

The first metric, PSD, serves to describe the distribution of signal power in the frequency domain and is quantified in units of power per Hertz. The preprocessed vibration segment $x(t)$ and the synthetic vibration segment $y'(t)$ were transformed into the frequency domain, represented as $X(f)$ and $Y'(f)$ respectively. This transformation is achieved by the Fourier transform. Subsequently, the PSD $S_{xx}(f)$ and $S_{y'y'}(f)$ are defined as the square of the magnitude of these frequency domain representations. This is expressed by the following equation:

$$S_{xx}(f) = |X(f)|^2 \tag{17}$$

$$S_{y'y'}(f) = |Y'(f)|^2 \tag{18}$$

where the raw signals are measured in units of gravity g, and the PSD is in units of g^2/Hz. This quantification allows a direct comparison of the power distribution of both signals over the frequency domain.

The second metric, the Modal Assurance Criterion (MAC), has traditionally been used in structural dynamics to assess the linear independence or correlation between mode shapes. The MAC measures the cosine of the angle between corresponding mode shape vectors at each frequency. The MAC was chosen to validate whether the vibration responses dominated by the bending modes of the wheelset in the raw signals are similarly implemented in the synthetic signals. After transforming the preprocessed vibration segment $x(t)$ and the synthetic vibration segment $y'(t)$ into the frequency domain via the Fourier transform, the MAC can serve as an indicator of the linear similarity between them. The MAC at frequency f is calculated as:

$$MAC(f) = \frac{|X(f) \cdot Y'(f)|^2}{|X(f)|^2 \cdot |Y'(f)|^2} \tag{19}$$

where $X(f)$ and $Y'(f)$ are the complex Fourier coefficients at frequency f. This results in a frequency-dependent measure of similarity, ranging from 0 to 1, where 0 indicates no correlation and 1 indicates perfect correlation. The overall similarity between both signals, across the entire frequency range, is quantified by the average of the MAC value, denoted as MAC_{avg}, which is computed as follows:

$$MAC_{avg} = \frac{1}{N} \sum_{f=1}^{N} MAC(f) \tag{20}$$

where N represents the total number of frequency bins derived from the Fourier transform.

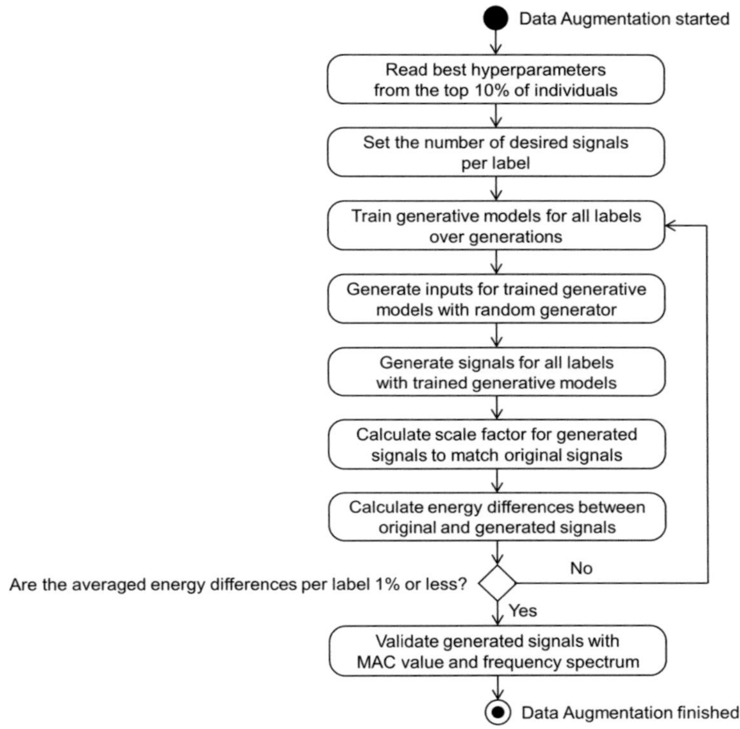

Figure 24: **Data Augmentation Process using LSTM-based Generative Model**

4.5 Results and Discussion

4.5.1 First Generation in Evolutionary Process

Table 12 shows the initial configurations of the evolutionary process for the LSTM network, listing the randomly distributed hyperparameters and their corresponding performance metrics for the first generation. These results were calculated based on 10 iterations of the same hyperparameters for generalisation. Figure 25 shows the comparison of the different synthetic vibration segments generated by these initial configurations at the first generation, categorised by their performance ranking: the best, top 25%, top 50%, and top 75% individuals out of 100 individuals.

For the best individual, the synthetic vibration data closely matches with the preprocessed vibration data across all four labels, indicating an initial model configuration that effectively captures the transient feature, caused by wheel flats, of the preprocessed vibration data during the evolutionary process. The top 25% individual shows a slight distortion in capturing the transient features of the preprocessed vibration data. The overall shape of the synthesised vibration data looks similar, but the centrally located transient feature looks somewhat smoother. The batch size and learning rate are different from the best individual. The slight difference is probably due to the small learning rate rather than the batch size. Therefore, the training was slower and then stopped before the training was complete. As a result, the training loss and fitness are a little poor, but similar values for the energy correction factor and MAC indicate that it has trained similarly with the best individual, except for the smooth transient feature.

For the top 50% and top 75% individuals, it is intuitively clear at a glance from the synthetic vibration data that the generative models could not adequately learn the preprocessed vibration data due to improper hyperparameter selection. This is likely due to the large number of LSTM cells compared to the best individual, rather than differences in batch size and learning rate. It is inferred that eight LSTM cells had a negative influence on the training process because their learning capacity was too large compared to the information of the preprocessed vibration data, which could be sufficiently learnt by one LSTM cell. The training loss, fitness, energy correction factor, and MAC show completely different values compared to the best individual. The large energy correction factor indicates that the output of the LSTM cells is very small, while the low MAC means that it is poorly correlated with the preprocessed vibration data.

Evolutionary Process	Last Generation	First Generation			
Rank	Best Individual	Best Individual	Top 25% Individual	Top 50% Individual	Top 75% Individual
Training Loss	0.000076	0.000250	0.009494	0.996231	0.996095
Fitness	0.999933	0.999647	0.989359	0.500944	0.500944
Parameter 1 Hidden State Dimension	32	16	16	16	16
Parameter 2 Number of LSTM Cell	1	1	1	8	8
Parameter 3 Batch Size	8	32	16	64	128
Parameter 4 Learning Rate	0.1	0.1	0.0001	0.1	0.001
Energy Correction Factor - Non-defect - Wheel Flat 1 - Wheel Flat 2 - Wheel Flat 3	0.661 0.886 0.889 0.851	0.645 0.846 0.723 0.956	0.672 0.908 0.911 0.878	156.138 73.207 97.727 92.093	34.317 708.977 488.332 87.468
MAC (Cosine Similarity) - Max - Mean - Min - Std	1.000 0.632 0.294 0.076	1.000 0.634 0.294 0.075	1.000 0.630 0.288 0.074	0.247 0.155 0.074 0.038	0.366 0.180 0.073 0.065
Processing Time - Mean - Std	6.565 0.652	217.911 13.677	76.401 7.641	25.023 4.069	23.821 3.394

Table 12:   **Comparison of Individuals and Their Performances at First and Last Generations**

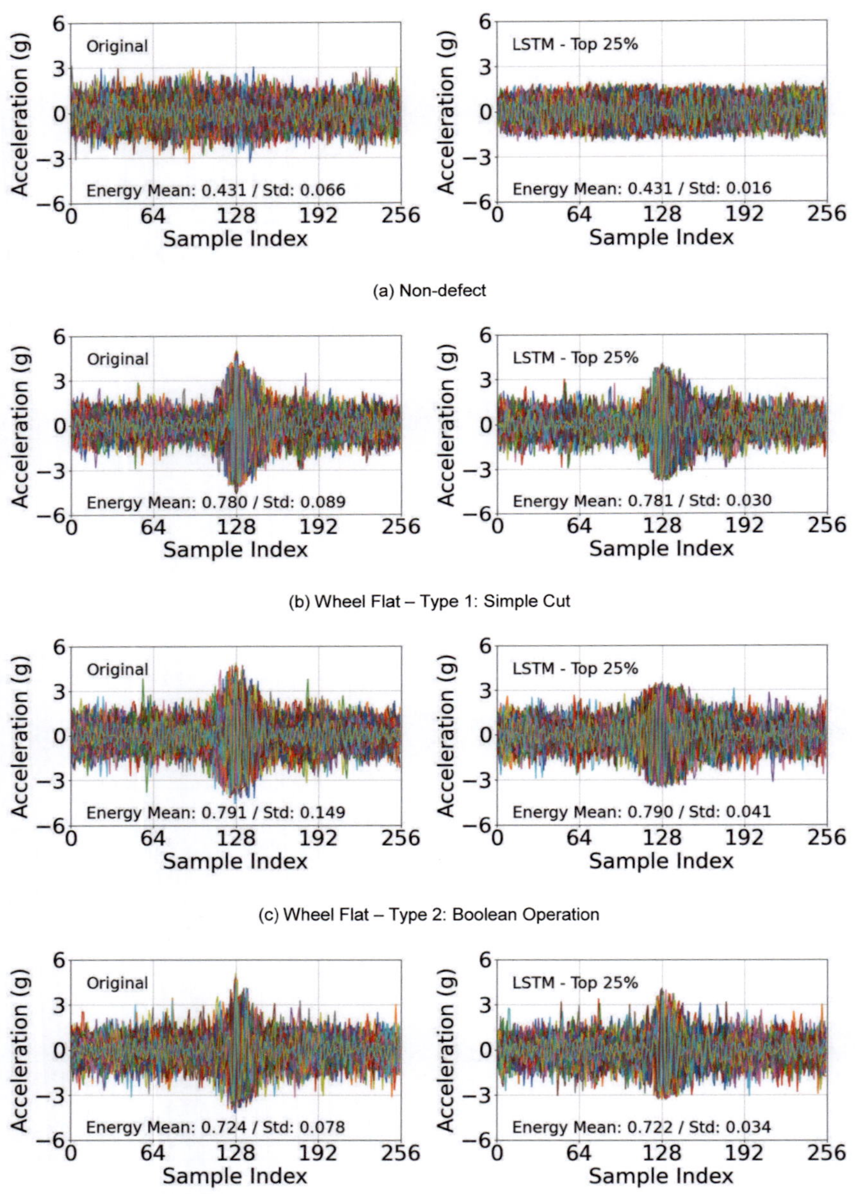

(a) Non-defect

(b) Wheel Flat – Type 1: Simple Cut

(c) Wheel Flat – Type 2: Boolean Operation

(d) Wheel Flat – Type 3: FEM Simulation

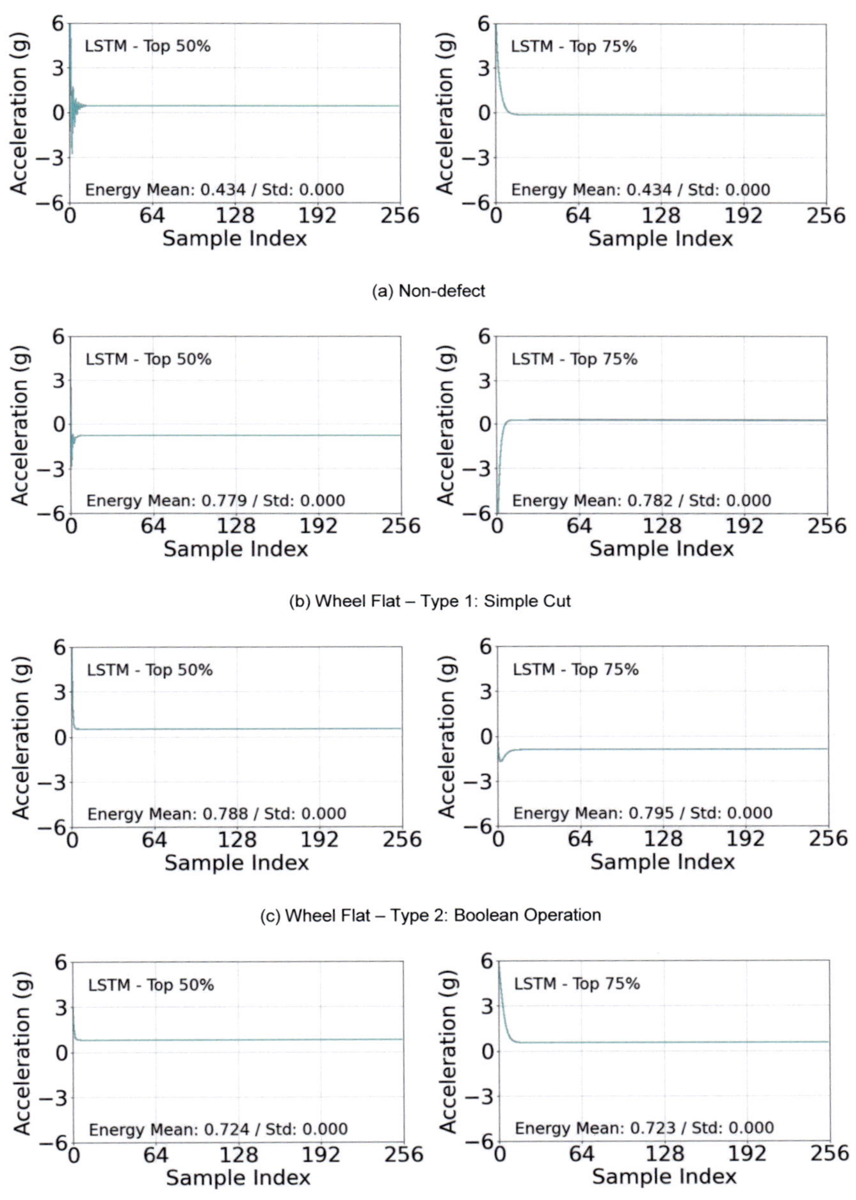

Figure 25: **Comparison of Synthetic Vibration Data by Label at First Generation**

4.5.2 Evolutionary Process and Termination

The diversity within a population is quantified by the Shannon entropy, which measures the variance of genetic information among the elite fraction of a population. The top 10 individuals out of a total of 100 individuals. Shannon entropy can be calculated using Equation (13). The genetic diversity is found to be greatest at the inception of the population, attributed to the random sampling of genetic information within a specified range for each hyperparameter. As the evolutionary process progresses, through a series of crossovers and mutations, the Shannon entropy decreases, indicating a consolidation of genetic information around a subset of the population. The process culminates when Shannon entropy reaches zero, which occurs in a range between the 8th and 16th generations.

Table 13 and Figure 26 detail the changes in Shannon entropy and training loss across generations, and Figure 27 shows the evolution of training loss, particularly in Trial 2. These figures represent the general trend observed across multiple trials. After the 4th generation, the evolutionary process is characterized by a diminishing, yet fluctuating, standard deviation in training loss until the end of the evolutionary process. This fluctuation is not a consistent feature across all evolutionary runs, but represents the pre-termination phase, which is influenced by the mutation rate determined to be optimal at 5% after trial and error. This mutation rate plays an important in preventing premature convergence to a particular genetic sequence, thus preserving genetic diversity that is essential for thoroughly exploring the hyperparameter space.

Table 14 shows Pearson correlation coefficients, quantifying the relationship between training loss and four hyperparameters throughout the evolutionary process. Parameter 2 (Number of LSTM cells). This means that the model's increased complexity leads to higher training loss. Meanwhile, Parameter 3 (Batch Size) displays a low correlation (0.429), implying that larger batch sizes may elevate training loss. In contrast, Parameter 4 (Learning Rate), shows an even lower correlation (0.270) than Parameter 3, with a modest rise in loss as the learning rate increases. Parameter 1 (Hidden state Dimension) shows an almost negligible correlation (0.033), indicating minimal influence on training loss. These insights highlight the importance of precise hyperparameter tuning, particularly the number of LSTM cells, which substantially affects training loss compared to others. These coefficients were calculated from a non-uniform distribution within the hyperparameter space, indicative of the GA's selective exploration, as opposed to a grid search's comprehensive evaluation. The overall positive correlation trend, except for the hidden state dimension, suggests a general propensity for increased hyperparameter values to adversely affect model performance, although the extent of this impact varies markedly among the parameters.

The convergence of hyperparameters during the evolutionary process is influenced by the GA's settings, such as population size, crossover rate, and mutation rate. Repeated evolutionary trials under the same conditions often leads to slightly different convergence of hyperparameters. For instance, Parameter 1 shows a tendency to converge to variable values, such as 16, which suggests a relatively minor role in influencing the objective function. This result was confirmed through visual analysis and is further discussed in the context of the results presented in Table 13.

The findings highlight the critical role of genetic diversity in the successful application of GA for hyperparameter optimisation, emphasising. the crucial interaction between hyperparameters and the evolutionary process in implementing the success of GA for hyperparameter optimisation.

Generation	1	2	3	4	5	6	7	8	9
Shannon Entropy	6.300	4.260	3.823	4.582	3.266	2.440	1.763	1.763	0.000
Training Loss	0.501 ±0.462	0.137 ±0.295	0.027 ±0.141	0.004 ±0.020	0.009 ±0.061	0.001 ±0.003	0.016 ±0.106	0.001 ±0.001	0.021 ±0.139

Table 13:   **Shannon Entropy and Training Loss across Generations**

Parameter Name	Parameter 1 Hidden State Dimension	Parameter 2 Number of LSTM Cell	Parameter 3 Batch Size	Parameter 4 Learning Rate
Training Loss	0.033	0.724	0.429	0.270

Table 14:   **Effects of Hyperparameters on Training Loss using Pearson Correlation Coefficients**

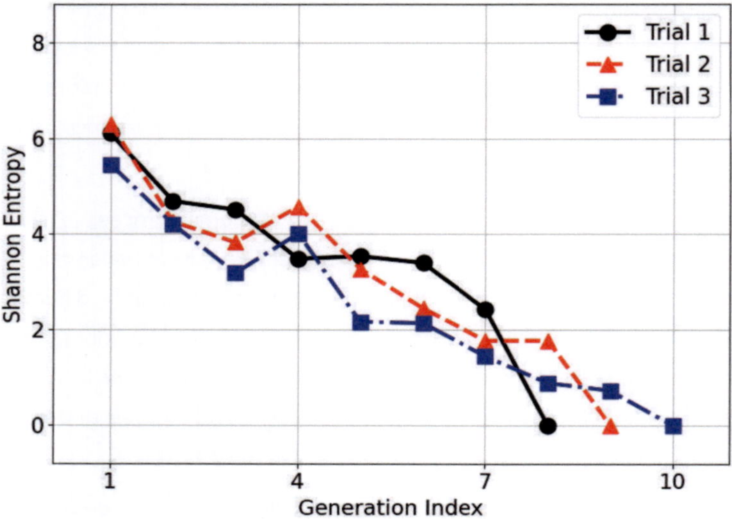

Figure 26:   **Shannon Entropy across Generations**

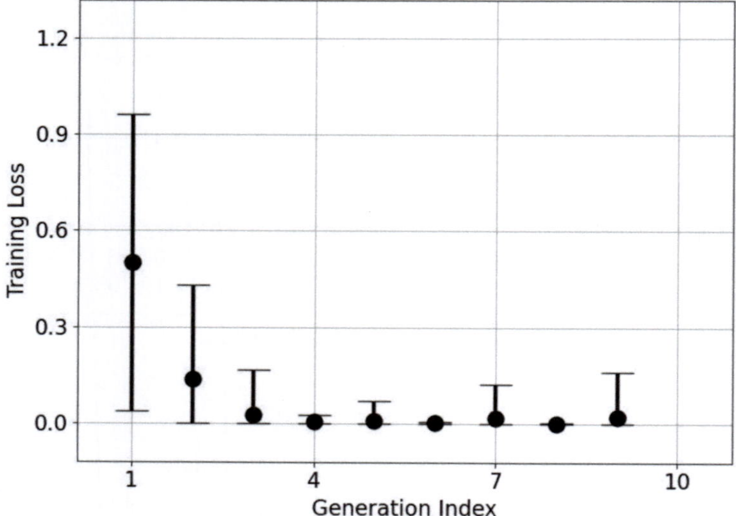

Figure 27:   **Training Loss across Generations from Trial 2**

4.5.3   Last Generation in Evolutionary Process

The GA converged on the best hyperparameters for the top 10% of individuals at the 9th generation, as shown in Table 12. The same 200 synthetic vibration segments per label were generated and compared with the preprocessed vibration data used as input to the LSTM-based generative algorithm. The results are as follows:

Figure 28 illustrates the comparison of preprocessed and synthetic vibration data by label at the last generation. Here, the vibration data is presented in terms of acceleration over a set of sample indexes for both preprocessed and LSTM predicted data across three different wheel flat types. Although energy mean values have been adjusted through scaling factors, thereby aligning the LSTM predictions closely with the preprocessed data, it is observed that the standard deviations of the LSTM predictions are somewhat smaller. This difference indicates that while the LSTM network approximates the central tendency of the data well, it tends to generate predictions with less variability than the preprocessed data.

Figure 29 continues the comparison by including the PSD and 95% confidence interval of the preprocessed and synthetic data by label. This plot is instrumental for identifying the frequency characteristics of the vibration data, with the mean and confidence intervals of the preprocessed and LSTM-generated data superimposed for comparison. It is observable that the LSTM generated data closely follows the trend of the preprocessed data, indicating effective learning by the model.

Figure 30 displays the MAC values in a matrix format for three cases. MAC is a statistical measure of the similarity between shapes, ranging from 0 to 1, where 1 indicates identical shapes. The matrices offer a visual comparison of signal similarity, with Case 1 comparing preprocessed data against itself, Case 2 comparing preprocessed data with synthetic data, and Case 3 comparing synthetic data against other synthetic data.

Figure 31 shows that the distribution of MAC values for the three cases is concentrated around MAC values between 0.6 and 0.7. The fact that even within the preprocessed data from Case 1, the average MAC value is between 0.6 and 0.7 shows that the data is not skewed in either direction but has diversity and provides a target point for the generative model. This indicates a strong similarity in the vibration pr-files of the three cases, which are almost identical. This distribution shows that the synthetic data generated by the LSTM network not only closely reflects the preprocessed data, but also reflects the consistency of the synthetic data itself, successfully synthesising the characteristic patterns present in the preprocessed dataset.

Table 15 and Table 16 complement the figures by offering a numerical distribution of MAC values and breaking down the statistical metrics of the MAC distributions for each case, respectively. These statistical measures provide a concise summary of the similarity assessments, offering a quantitative method to gauge the effectiveness of the LSTM network in synthesising the vibration characteristics of the preprocessed data.

Table 12 shows that while both the best hyperparameters of the first and last generations were able to generate synthetic vibration data that closely resembled the preprocessed vibration data, there is a notable difference in the complexity of the implemented LSTM network. This is evidenced by the significant reduction in processing time from the first to the last generation. Initially, the best individual model from the first generation, with a hidden state dimension of 16 and a batch size of 32, took an average of 217.911 seconds to generate 200 synthetic vibration segments per label. However, in the latest generation, the best model, with a more complex structure, a hidden state dimension of 32 and a smaller batch size of 8, was able to produce these segments in just 6.565 seconds on average.

This significant improvement in the process time shows that not only has the fidelity of the synthetic vibration data been maintained over the evolutionary process, but the computational efficiency of the LSTM network has also improved significantly. The optimisation of the hyperparameters has resulted in a LSTM network that is not only more accurate but also faster, indicating an optimisation in both quality of output and operational efficiency.

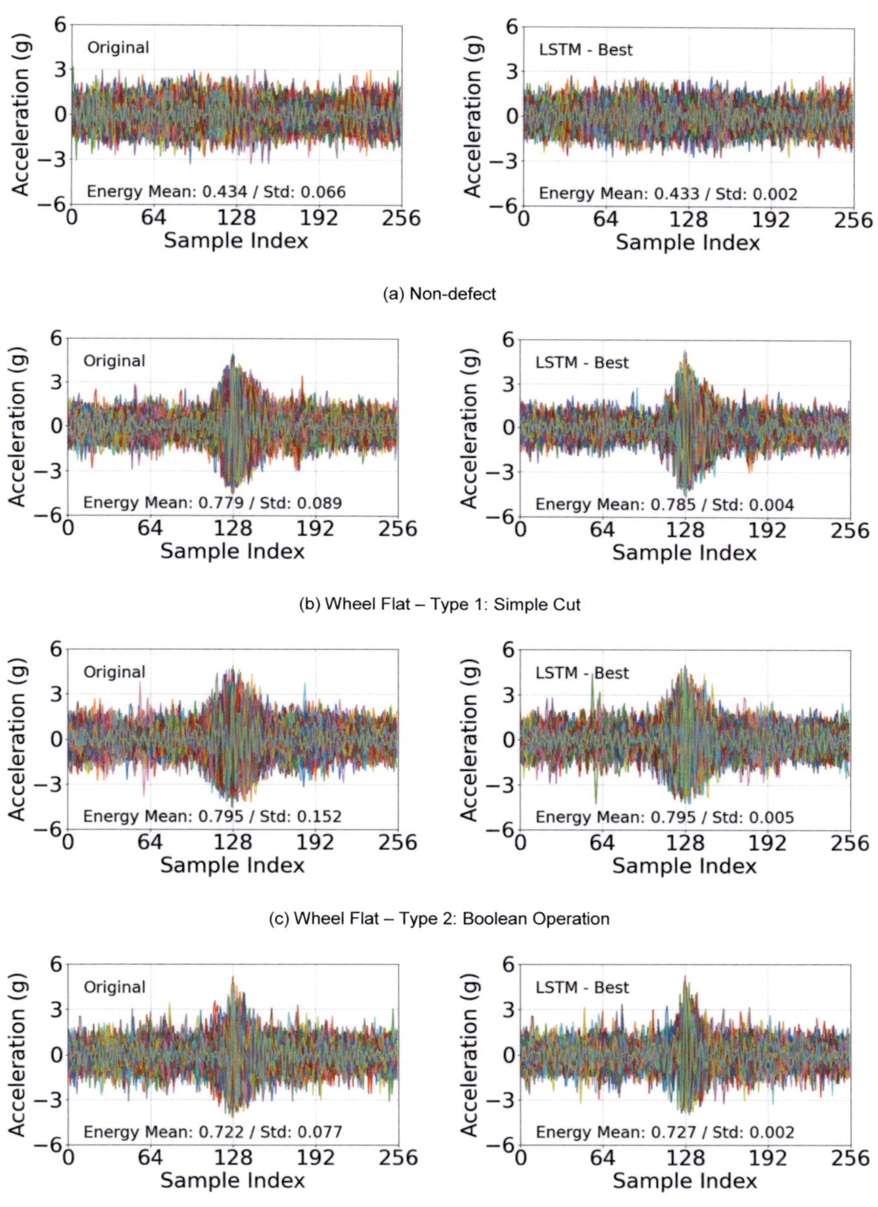

(a) Non-defect

(b) Wheel Flat – Type 1: Simple Cut

(c) Wheel Flat – Type 2: Boolean Operation

(d) Wheel Flat – Type 3: FEM Simulation

Figure 28:   **Comparison of Preprocessed and Synthetic Vibration Data by Label at Last Generation**

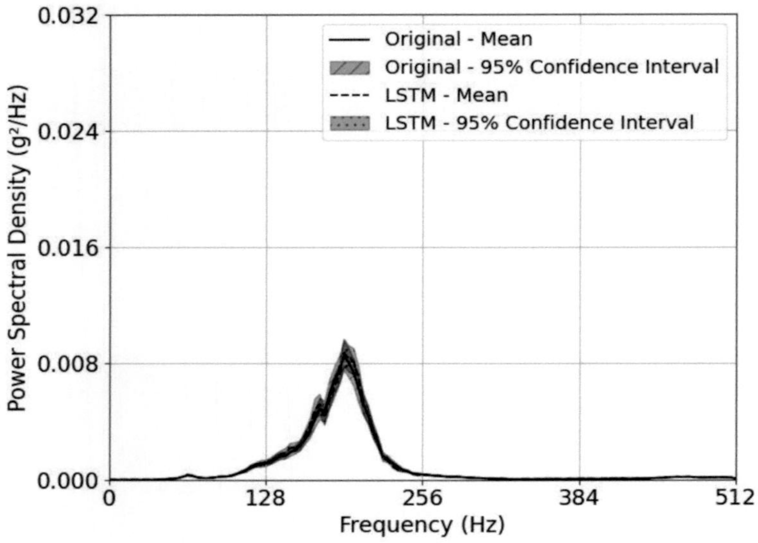

(a) Non-defect

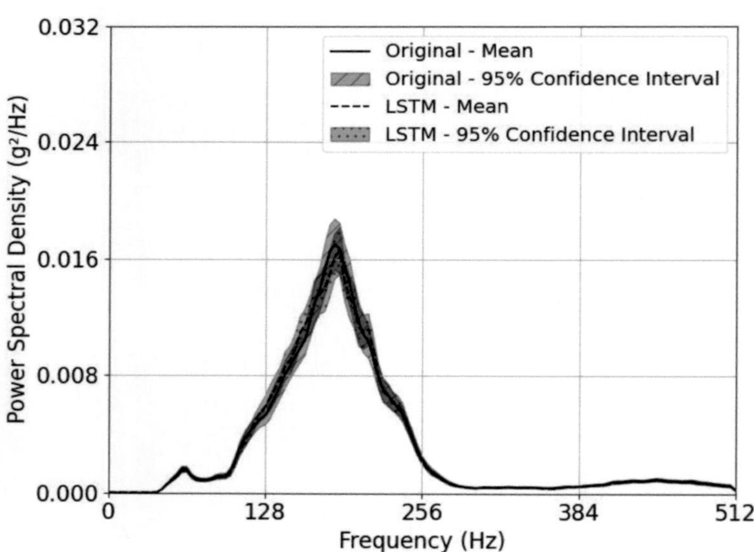

(b) Wheel Flat – Type 1: Simple Cut

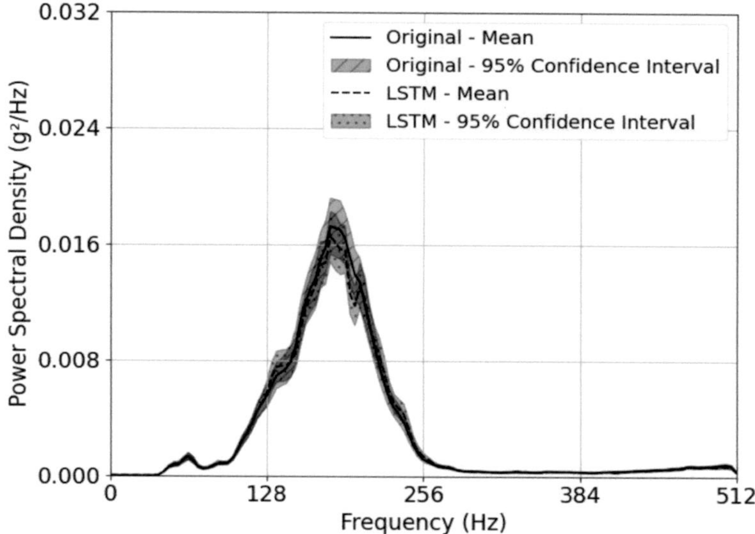

(c) Wheel Flat – Type 2: Boolean Operation

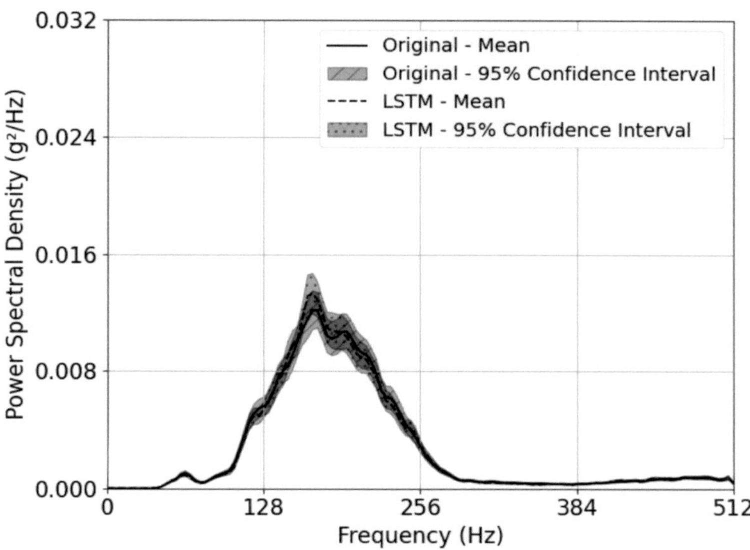

(d) Wheel Flat – Type 3: FEM Simulation

Figure 29: **Comparison of Power Spectral Density and 95% Confidence Interval of Preprocessed and Synthetic Data by Label**

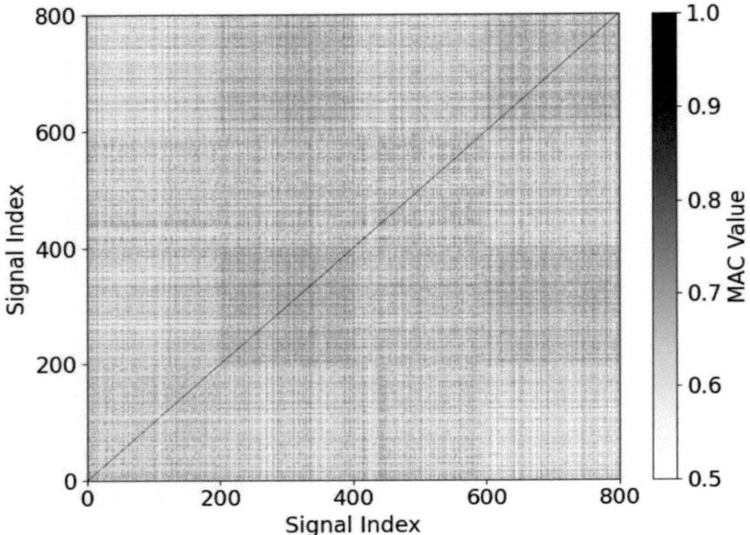

(a) Case 1: Preprocessed Data versus Preprocessed Data

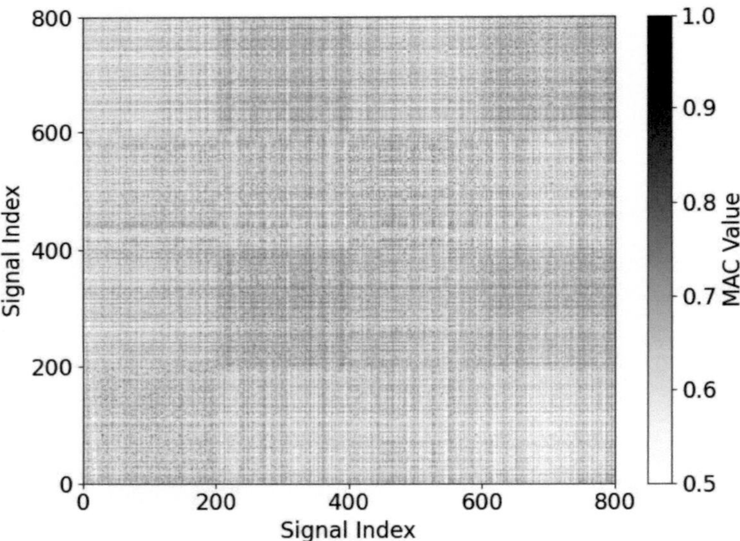

(b) Case 2: Preprocessed Data versus Synthetic Data

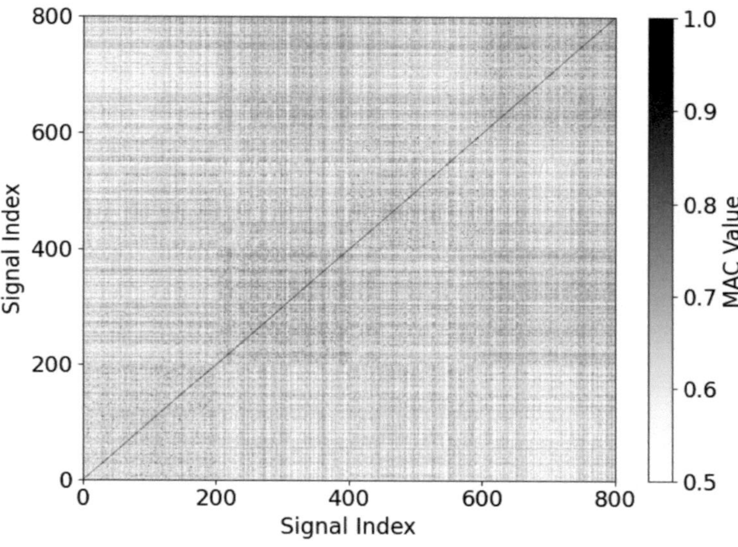

(c) Case 3: Synthetic Data versus Synthetic Data

Figure 30: **Comparison of Signal Similarity using Modal Assurance Criterion by Case**

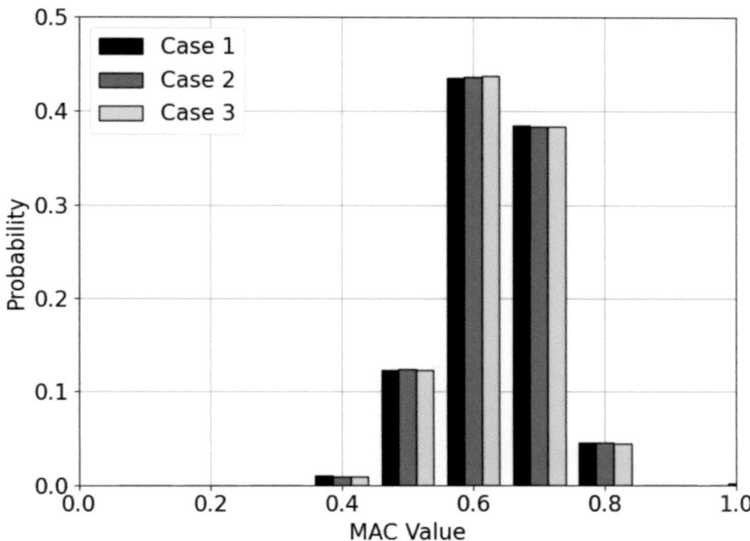

Figure 31: **Distributions of Modal Assurance Criterion by Case**

Midpoint of Bins	0.1	0.2	0.3	0.4	0.5	0.6	0.7	0.8	0.9
Case 1	0.000	0.000	0.000	0.010	0.122	0.434	0.388	0.046	0.000
Case 2	0.000	0.000	0.000	0.010	0.122	0.431	0.389	0.046	0.000
Case 3	0.000	0.000	0.000	0.011	0.121	0.429	0.390	0.047	0.000

Table 15: **Modal Assurance Criterion Distributions by Case**

Statistical Metrics	Max	Mean	Min	Std
Case 1	0.999	0.633	0.295	0.074
Case 2	1.000	0.632	0.294	0.076
Case 3	1.000	0.631	0.295	0.076

Table 16: **Statistics of Modal Assurance Criterion Distributions by Case**

5 Enhanced Detection of Different Wheel Flat Geometries using Augmented Vibration Data

This chapter aims to explore how to combine the synthetic vibration data with the pre-processed vibration data and which wheel flat modelling method is suitable for follow-up studies. The tasks are detailed as follows: 1) preparing datasets for case studies, 2) modelling, training, and validating a CNN-based classification model, 3) optimizing the hyperparameters of the classification model using the grid search algorithm, and 4) discussing the influence of above three aspects on classification performance.

5.1 Dataset Preparation for Case Studies

In the previous two chapters, the preprocessed vibration data were obtained physically from the 3D-printed scale model, while the synthetic vibration data were produced digitally using the LSTM-based generative model. In this subchapter, the case studies were conducted to explore how to effectively prepare the input datasets to minimise the training loss of the classification model while increasing its validation and testing accuracy. Figure 32 shows three scenarios based on the different combination of data and the different proportions of data for training, validation, and testing.

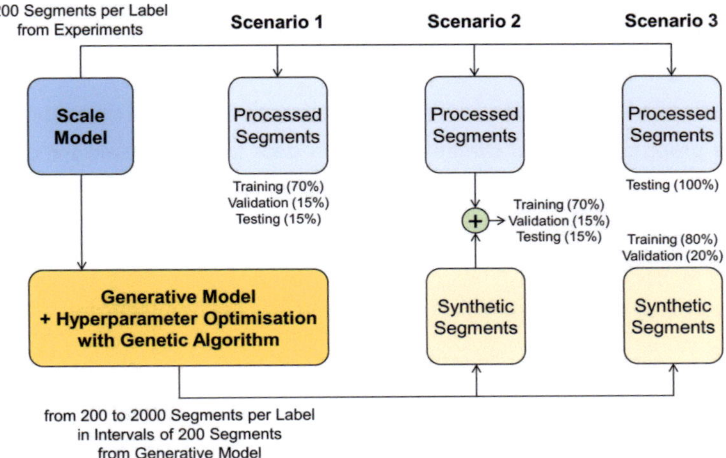

Figure 32: **Dataset Schemes for Case Studies**

The raw vibration data obtained through dynamic testing on the 3D-printed scale model was segmented and filtered through preprocessing. As a result, the preprocessed vibration data consisted of 200 segments per label and a total of 800 segments for the four labels were prepared for Scenario 1. The segments were then divided into three datasets: training (70%), validation (15%), and testing (15%). This is common practice in deep learning models to prevent underfitting or overfitting during training.

Similar to performing a field experiment with a real railway vehicle and track, conducting a dynamic testing with the 3D-printed scale model in the laboratory also demands significant time and effort. Therefore, it is still considered difficult to experimentally obtain the raw vibration data required to train a classification model, even from the 3D-printed scale model. Alternatively, Scenarios 2 and 3 use commonly both the preprocessed and the synthetic vibration data together. The difference between both scenarios starts with the first question of how to augment vibration data effectively. Scenario 2 combines both the preprocessed and synthetic vibration data and divides it into datasets for training, validation, and testing in the same proportions as Scenario 1.

In contrast, Scenario 3 uses only the synthetic vibration data for training and validation. The synthetic vibration data was divided into datasets for training and validation in the proportion of 80% and 20%, respectively. And only the preprocessed vibration data was used for testing. From the results in Chapter 4, it was concluded that the synthetic vibration data looks similar to the preprocessed vibration data, but there are still small differences. Among the 3V (volume, variety, and velocity), the synthetic vibration data can fill the gap in volume and velocity, but it is questionable whether it has enough variety independently. If it does not have enough variety, it would be better to combine both the preprocessed and synthetic vibration data to compensate for each other's weaknesses, as in Scenario 2. On the other hand, if the synthetic vibration data has enough variety, it can be properly trained without the preprocessed vibration data during training and validation, as in Scenario 3. This will be verified by comparing the testing accuracy obtained from evaluations using only the preprocessed vibration data.

The next question is how much synthetic vibration data should be added to help training, validate, and testing the classification model. The amount of synthetic vibration data will be increased from 200 to 2,000 per label, in increments of 200, to evaluate its effectiveness at 1x to 10x the amount of the preprocessed vibration data per label.

5.2   Modelling of Detection Algorithm using Convolutional Neural Network

Convolutional Neural Networks (CNNs) are a deep learning architecture that excels at processing data with a grid-like topology, such as images. CNNs use convolutional layers to automatically detect and learn a spatial hierarchy of features, without the need for manual feature extraction. This capability makes CNNs very effective for tasks like image recognition and object detection. This has been used in many existing studies on the detection of wheel flats. In this study, the automatic feature extraction capability of CNNs was exploited by using 1D convolutional layers. These layers are capable of extracting features from the pewprocessed vibration data, which is presented in the form of 1D time-series data, and then classifying these features into one of four labels. Figure 33 shows the architecture of the 1D CNN model that was finally selected through trial and error. From input layer to output layer, the details are as follows:

In the preprocessed vibration data, the length of each segment is 256 samples. These segments were initially preprocessed with nomalisation and Fast-Fourier Transform (FFT) before feeding them into the 1D convolutional neural network. In the input layer, the normalisation is a common technique for adjusting input values to a standard range, which facilitates more stable and faster convergence during training.

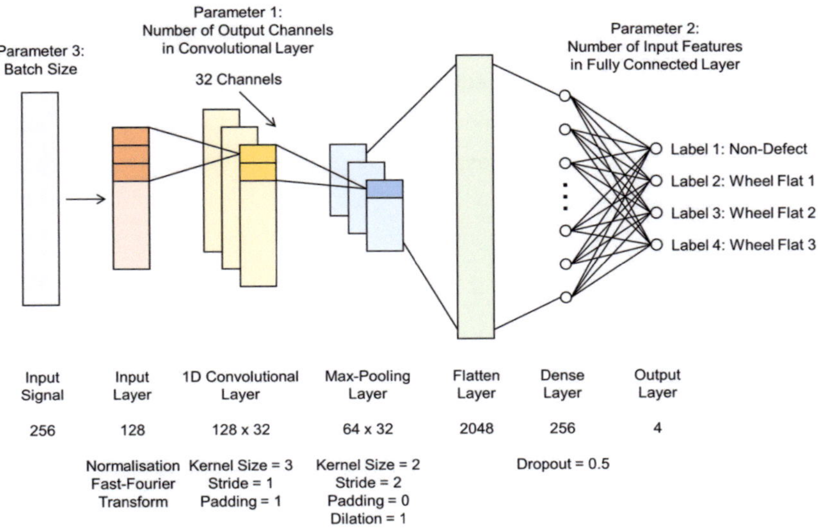

Figure 33:   **Architecture of 1D Convolutional Neural Network (CNN) Model**

The normalised segments were subsequently transformed from real values in the time domain to complex values in the frequency domain using FFT. The energy normalisation was performed by dividing the FFT output by the number of samples, 256, and multiplying by 2 to account for the symmetric property of the spectrum. Only the front half of the FFT output is retained due to this symmetry. The FFT operation is as follows:

$$X(k) = \sum_{n=0}^{N-1} x(n) \cdot e^{-\frac{i2k}{N}kn} \tag{21}$$

where $X(k)$ is the FFT output of time-series data $x(n)$, N is the number of samples in the segment, specified as 256, and i is the imaginary unit. The FFT output $X(k)$ is then converted from complex values to the spectrum with absolute values:

$$|X(k)| = \sqrt{\Re\big(X(k)\big)^2 + \Im\big(X(k)\big)^2} \quad for \quad k = 0 \; to \; 127 \tag{22}$$

where $|X(k)|$ represents the absolute spectrum, $\Re\big(X(k)\big)$ is the real part, and $\Im\big(X(k)\big)$ is the imaginary part of the complex spectrum $X(k)$. The convolutional operation applies 1D filters to this absolute spectrum of 128 samples in length to produce feature maps. The parameters of this operation are as follow: The kernel size set to 3 and defines the width of the filter that is used to capture local patterns within the absolute spectrum. The stride set to 1 and determines the step size the filter takes as it moves across the absolute spectrum, influencing the resolution of the feature maps. The padding set to 1 and indicates the amount of zero-padding added around the absolute spectrum to allow 1D filters to process the edges without information loss. This convolutional operation is defined by:

$$f_i(x) = \sum_{j=1}^{3} k_{i,j} \cdot |X(s + j - 2)| \quad for \quad i = 1 \; to \; 128 \tag{23}$$

where $f_i(x)$ represents the output of the convolutional operation at position i within the absolution spectrum, $\sum_{j=1}^{3}$ is a summation loop over the three elements of the kernel filter, $k_{i,j}$ is the three coefficients of the kernel filter at position j within the kernel for position i, $|X(s + j - 2)|$ is the absolute value of the absolute spectrum at a specific position, where s is the starting position of the convolution windows. This convolution operation, through the summation of the products of the kernel coefficients and the absolute spectrum, effectively extracts features from the absolute spectrum. As the filter slides across the absolute spectrum with the specified stride and padding, it applies the weights of the kernel to the input values, producing a transformed output. This output is a feature map that represents localised features of the absolute spectrum. By capturing the essential characteristics of the absolute spectrum within these feature maps, the convolutional layer can detect patterns and details that are critical for the subsequent layers to perform further analysis or classification tasks.

Following the convolutional layer, the max-pooling layer serves to condense the feature map by reducing its spatial dimensions. This process emphasises the most significant features while simplifying the data volume. The max-pooling operation selects the maximum value within a certain region, defined by the kernel sliding over the feature map obtained from the convolutional layer. The max-pooling operation can be described as:

$$p_i(x) = max_{1 \leq j \leq 2} \left(f_{2i+j-2}(x) \right) \quad for \quad i = 1 \text{ to } 64 \tag{24}$$

where $p_i(x)$ is the output of the max-pooling operation, $max_{1 \leq j \leq 2}$ is the maximisation operation over two elements within the pooling window, $f_{2i+j-2}(x)$ is the feature map obtained from the convolutional layer. The index $2i + j - 2$ is a specific position within the pooling window, determined by the pooling stride and the size of the kernel. During the max pooling process, the feature map of 128 samples in length halves to 64 sample in length. Therefore, i is from 1 to 64.

After the max-pooling layer, the flattening layer transforms the multi-dimensional pooled feature maps into a one-dimensional vector. This transformation is necessary for transitioning from the spatially structured output of the convolutional layer to the format required by the dense layer. The flattening operation can be represented as:

$$v = flatten\big(p(x)\big) \tag{25}$$

where v is the one-dimensional vector that contains all the elements of the pooled feature map $p(x)$ and $flatten$ function takes the output tensor of the max-pooling layer and restructures it into the vector v.

The dense layer in a neural network facilitates high-level reasoning by forming non-linear combinations of features extracted by the convolutional layers. Each neuron in the dense layer is connected to all activations from the previous layer, enabling the network to integrate these features into more abstract representations. The operation within the dense layer is defined as below:

$$z = ReLU(W \cdot v + b) \tag{26}$$

Where z is the activation vector produced by the dense layer, $ReLU$ represents the Rectified Linear Unit, which is the non-linear activation function and allow it to learn more complex patterns, W is the weight matrix, v is the input vector from the flattening layer, b is the bias vector.

To mitigate the risk of overfitting, a dropout technique is employed at this layer with a rate of 0.5. This means that during training, each neuron has a 50% chance of being omitted from the layer's calculations, effectively reducing the model's complexity and promoting more robust learning. The dropout operation can be represented:

$$z' = m \odot z \tag{27}$$

Where z represents the new output after applying dropout and m is a random binary mask (0 or 1) with a probability corresponding to the dropout rate.

Finally, the output layer consists of four neurons corresponding to the four labels. This layer employs a softmax activation function to convert the output of the neurons into a probability distribution over the labels. The sum of the probabilities for all labels equals 1. The softmax function is as follows:

$$\sigma(z_i') = \frac{e^{z_i'}}{\sum_{j=1}^{4} e^{z_j'}} \quad for \quad i = 1\ to\ 4 \tag{28}$$

where $\sigma(z_i')$ denotes the probability that the given input is classified into the i-th label, z_i' is the input to the softmax function from i-th neuron in the output layer, and $e^{z_j'}$ is the exponential transformation of the i-th input, ensuring all outputs are ono-negative, $\sum_{j=1}^{4} e^{z_j'}$ is the sum of the exponential transformations of all neuron inputs in the output layer, acting as a normalising constant.

5.3 Hyperparameter Optimisation using Grid Search Algorithm

5.3.1 Grid Search Algorithm

The choice of hyperparameters in training a classification model is critical to its classification performance. In this subchapter, the Grid Search algorithm, which is simpler and more intuitive rather than GA, was used to find the top 5 hyperparameter combinations based on training loss. Figure 34 shows the hyperparameter optimisation process. Low training loss does not always guarantee good results. Therefore, the training, validation, and testing process was further validated to find the best hyperparameter combination within the top 5 hyperparameter combinations. The classification model was implemented based on the CNN model provided by the 'nn.Module' in the PyTorch framework, and then was trained using a NVIDIA GeForce RTX 2060 Super on a Window 10 system, utilizing Python 3.9 and PyTorch 2.1.

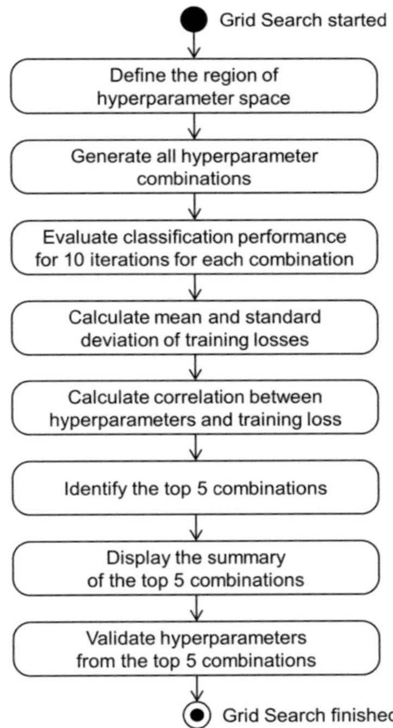

Figure 34: **Hyperparameter Optimisation Process using Grid Search Algorithm**

5.3.2 Selection and Range of Hyperparameters

The hyperparameter optimisation begins with defining the range of the hyperparameter space. Since there is no prior information for this process, the appropriate hyperparameters and their ranges were determined through trial and error in this study, as detailed in Table 17. A single convolution layer was chosen to be sufficient because of the intuitive distinction in the vibration data. Four hyperparameters were selected that are estimated to have a high contribution to classification performance. The description of each hyperparameter is as follows:

- **Parameter 1 - Quantity of Output Channels in Convolutional Layer**: Determines the depth of the feature map, allowing for the detection of diverse features and affecting the network's complexity and the computational load.

- **Parameter 2 - Quantity of Input Features in Fully Connected Layer**: Determines the size of the dense layer, which correlates with the network's capacity for integrating and interpreting extracted features, affecting both the learning capability and the computational load.

- **Parameter 3 - Batch Size**: Affects the number of segments processed in one update, affecting the network's learning stability and efficiency. Smaller sizes can improve generalisation, while larger ones can enhance gradient accuracy.

- **Parameter 4 - Learning Rate**: Control the learning speed, affecting the network's convergence rate. If set it too high, it may skip the optimal result, while too low a rate might result in slow convergence.

5.3.3 Hyperparameter Optimisation

Among the various search algorithms, the grid search algorithm was used to find the best hyperparameter combination by sequentially exploring all hyperparameter combinations. For more efficient computation, alternatives such as randomised search and Bayesian optimisation can also be considered. Random search accelerates the search process by stochastically sampling the hyperparameter space, while Bayesian optimisation uses a more sophisticated probabilistic model that uses past evaluation data to efficiently narrow down the most promising hyperparameter combinations. Among the various search algorithms, the Grid Search algorithm was used to sequentially explore all combinations of hyperparameters to find the best hyperparameter combination. In this study, the Grid Search was chosen because it is intuitive and simple compared to other methods, the number of all the possible hyperparameter combinations is not large, and the computational demand for each hyperparameter combination is not excessively high.

In Table 17, each hyperparameter has five possible values, leading to a total number of possible hyperparameter combinations is 625. Repeating training on the same hyperparameter combination produced slightly different results. This is a common phenomenon and results in inconsistency in the top 5 hyperparameter combinations when sorting in ascending order with training loss after training on all hyperparameter combinations. Therefore, for each hyperparameter combination, training was performed 10 times instead of once. The mean and standard deviation of the training losses were calculated. Computing all hyperparameter combinations 10 times takes about 3 hours on a single GPU (NVIDIA GeForce RTX 2060 Super). The hyperparameter combinations were then consistently sorted in ascending order with the average training loss across repeated runs.

Parameter Name	Parameter 1 Quantity of Output Channels in Convolutional Layer	Parameter 2 Quantity of Input Features in Fully Connected Layer	Parameter 3 Batch Size	Parameter 4 Learning Rate
Possible Value	16 32 64 128 256	16 32 64 128 256	8 16 32 64 128	1 0.1 0.01 0.001 0.0001

Table 17:   **Range of Hyperparameter Space for Grid Search Algorithm**

5.4 Results and Discussion on Classification Performance

5.4.1 Effects of Hyperpameters

As shown in Table 18, exploring the hyperparameter space with the Grid Search algorithm results in an optimal set of configurations that minimise the average training loss. Particularly, Parameter 2 (Quantity of Input Features in Fully Connected Layer) and Parameter 4 (Learning Rate) tended to converge to 256 and 0.001 in the best performing hyperparameter set. All the top combinations share the value 256 for the quantity of input features value and 0.001 for the learning rate, suggesting that these parameters have a significant influence on classification performance. As shown in Table 17, the convergence of Parameter 2 to the upper limit (256) of its range highlights the importance of the size of the dense layer in the network's capacity to learn and generalise from the data while maintaining computational efficiency.

The variability in the top combinations for Parameter 1 (Quantity of Output Channels in Convolutional Layer) and Parameter 3 (Batch Size) indicates their influence on the training loss is less pronounced than Parameters 2 and 4. The best combination of hyperparameters, which results in the lowest training loss, employs 32 output channels and a batch size of 16. This suggests a balanced model complexity that enhances learning without unnecessarily increasing computational demand. Correlation analysis, as depicted in Table 19, further quantifies the relationship between hyperparameters and training loss. Parameter 4 demonstrates a positive correlation coefficient of 0.660 with the training loss, implying a critical role in the convergence behaviour. The training loss can be significantly reduced by fine-tuning the learning rate, which is evidence that the learning rate plays an important role in the training process.

Conversely, the other parameters exhibit negligible correlations. These includes a slightly positive correlation for Parameter 1 and a minor negative correlation for Parameter 2, which may suggest that the influence of these parameters on the classification performance does not follow a simple linear trend and that other non-linear relationships may exist. It is important to note the limitations of the linear correlation analysis due to the non-uniform distribution of hyperparameters, as indicated by the preset bounds in Table 17. While the grid search provides a valuable starting point, the implications of the Pearson correlation coefficients require further investigation of non-linear relationships within a more diversely distributed hyperparameter space.

Rank	Average Training Loss	Parameter 1 Quantity of Output Channels in Convolutional Layer	Parameter 2 Quantity of Input Features in Fully Connected Layer	Parameter 3 Batch Size	Parameter 4 Learning Rate
1 (Best)	0.111 ± 0.033	32	256	16	0.001
2	0.116 ± 0.042	32	256	32	0.001
3	0.122 ± 0.016	64	256	32	0.001
4	0.126 ± 0.031	16	256	16	0.001
5	0.131 ± 0.049	64	256	16	0.001

Table 18: Top 5 Hyperparameter Combinations searched by Grid Search Algorithm

Parameter Name	Parameter 1 Quantity of Output Channels in Convolutional Layer	Parameter 2 Quantity of Input Features in Fully Connected Layer	Parameter 3 Batch Size	Parameter 4 Learning Rate
Average Training Loss	0.130	-0.086	0.072	0.660

Table 19: Effects of Hyperparameters on Average Training Loss

5.4.2 Effects of Data Augmentation

5.4.2.1 Scenario 1 – only with Preprocessed Vibration Data

The classification performance of the model using only preprocessed vibration data serves as a baseline for understanding its capabilities without the enhancement provided by synthetic vibration data. These results are based on statistical values obtained after training the classification model using the best hyperparameters for 100 iterations. Otherwise, the scatter in the results from each training makes comparison and analysis somewhat difficult.

According to the data in Table 20, the classification model shows a training loss of 0.169 and a validation loss of 0.550 with a relatively modest data set of 200 segments per label. Despite these losses, the classification model shows a validation accuracy of 80.6% and a testing accuracy of 80.7%. While neither accuracy is higher than 95%, the nearly equal accuracy in validation and testing indicates that the classification model has some generalisation based on preprocessed real vibration data. Both low accuracies are likely due to underfitting due to lack of data, indicating that more data is needed to train, validate, and test the classification model.

The confusion matrix for Scenario 1, as shown in Table 23 (a), indicates a high true positive rate for label 1. However, it also indicates potential improvements in the discrimination of the remaining labels, as evidenced by the non-zero entries outside the main diagonal. This confusion in the classification of labels 2, 3 and 4 may be due to the limited quantity of the preprocessed real vibration data and the high similarity between them in terms of signal shape.

Figure 39 (a) shows a rapid decrease in the training loss with an initially steep learning curve as the model is trained over successive epochs. However, the flattening of the validation loss curve over epochs suggests that the classification model could benefit from an expanded dataset or additional hyperparameter optimisation to improve its learning efficiency and avoid underfitting. In addition, the accuracy trend in Figure 40 (a) for scenario 1 shows some variability across epochs. This variability highlights the possibility that the performance of the classification model could be further improved with a larger vibration data. Such variability is indicative of the challenges the classification model faces in capturing the complexity of the problem space with the limited amount of preprocessed vibration data on which it has been trained.

5.4.2.2 Scenario 2 – with Combination of Preprocessed and Synthetic Vibration Data

Table 21 presents the classification performance for Scenario 2, showing the incremental improvement in the model performance after adding synthetic vibration data, which varies from 200 to 2,000 segments per label at intervals of 200 segments. These results are also based on statistical values derived from 100 iterative trainings for each hyperparameter combination as in Scenario 1.

A consistent decrease in the training loss from 0.065 to 0.021 is observed as the number of synthetic vibration data increases, indicating better learning efficiency. Similarly, the validation loss decreases from 0.315 to 0.046, indicating that more synthetic vibration data improves the generalisation ability of the classification model. For both validation and testing, the accuracy increases with the number of synthetic vibration data. The validation accuracy starts at 0.911 for 200 segments and increases to 0.988 for 2,000 segments. The testing accuracy follows a similar upward trend from 0.910 to 0.989. These improvements indicate that adding synthetic vibration data significantly increases the prediction accuracy of the classification model.

The trends highlighted in Table 21 are also shown in the graphs in Figures 35 and 37. Figure 35 (a) shows a downward trend for both the training and validation losses, with the validation loss initially decreasing more steeply before flattening out, indicating diminishing returns beyond a certain amount of synthetic vibration data. Figure 35 (b) shows an increase in both the validation and testing accuracy as more synthetic vibration data is added, demonstrating the beneficial effect of data augmentation on the prediction accuracy of the classification model.

Figure 37 represents the effect of adding synthetic vibration data during the training process for Scenario 2. The slopes of both the loss and accuracy metrics initially show a large decrease. However, as more synthetic vibration data is added, the loss and accuracy slopes tend to decrease significantly, eventually the slopes become almost flat. This means that there is a limit to the advantageous of adding synthetic vibration data. The threshold at 5% of maximum loss or accuracy was drawn as an arbitrary reference line, not based on any specific standard or criteria. The improvement in slopes appears to plateau in the region of adding around 800-1,200 segments per label. During the training process, the optimal number of segments was determined to be around 1,200 segments after comparing several cases in terms of the training stability in the training loss and accuracy curves. The following figures are based on the results of adding 1,200 segments per label.

Figure 39 (b) compares the training and validation losses over different epochs in Scenario 2, showing a rapid adaptation of the classification model to the vibration data patterns, as the training loss drops sharply and then stabilises. The validation loss follows the same pattern, indicating good generalisation of the classification model without underfitting.

Figure 40 (b) shows the validation accuracy over several epochs for Scenario 2, which remains stable after an initial increase. This stability indicates a robust model that performs consistently well on the validation set, suggesting that the amount of synthetic vibration data used for training is sufficient to maintain high prediction performance without the need for additional synthetic vibration data.

5.4.2.3 Scenario 3 – with Synthetic Vibration Data for Training and Validation + with Preprocessed Vibration Data for Testing

Table 22 details the classification performance for Scenario 3, revealing the incremental effects on the model performance when trained only on synthetic vibration data, ranging from 200 to 2,000 segments per label at intervals of 200 segments. These results are also based on statistical values derived from 100 iterative trainings for each hyperparameter combination as in Scenario 2.

As the amount of synthetic vibration data gradually increases, the training loss steadily decreases from 0.019 to 0.010, indicating that the capability of the classification model to learn from synthetic vibration data is improving. The validation loss also follows this trend, decreasing from 0.174 to 0.001. The decrease in both losses indicates that adding synthetic vibration data helps the classification model generalise. However, when comparing the validation and testing accuracies, there is a different trend from Scenario 2. The validation accuracy starts at a high of 0.950 with just 200 segments per label and improves to a perfect accuracy of 1.000 already in 1,600 segments per label. In contrast, the testing accuracy shows practically no increase. This means that adding synthetic vibration data does not improve the trained classification model's ability to generalise to unseen data.

Figure 36 (a) for Scenario 3 highlights a sharp decline in both training and validation losses initially, similar to what we see in Scenario 2. However, the validation loss plateaus sooner, hinting that the model may not require as much synthetic data to reach its performance threshold. Figure 36 (b) confirms that both validation and testing accuracies remain high as more synthetic data is incorporated, but these metrics plateau relatively quickly. This suggests that adding synthetic data beyond a certain threshold does not significantly push the model's accuracy further.

The rate of change in the model's performance, as shown in Figure 38 for Scenario 3, begins to level off after an initial period of significant gains, much like in Scenario 2. This levelling indicates that the model extracts maximal learning from the synthetic data early on, and additional data doesn't yield proportionate benefits. The learning behaviour over multiple epochs, demonstrated in Figure 39 (c) for Scenario 3, shows the model quickly grasping the patterns within the synthetic data, as reflected by the rapid decline in training loss. The validation loss not only decreases sharply but also stays consistently low, suggesting that Scenario 3's synthetic data leads to robust generalisation. Figure 40 (c)'s portrayal of validation accuracy over successive epochs for Scenario 3 shows a model that maintains a high level of accuracy consistently, without the fluctuations seen in Scenario 2. This implies that the synthetic data is adequate for training a well-tuned model capable of sustaining high predictive performance without further data additions. From these insights, it appears that Scenario 3's model efficiently utilises synthetic data, achieving optimal performance with a moderate amount of data, likely between 800 to 1,200 segments, beyond which the returns diminish.

5.4.2.4 Comparison between Scenarios 1, 2, and 3

Scenario 1, where only the preprocessed vibration data segments from the scale model were used, is the baseline for comparison with the other two scenarios where synthetic vibration data is added in different ways. As shown in Table 20, the losses and accuracies during training, validation, and testing are worse or similar to the gradual improvements in the other two scenarios.

Scenario 2 aims to determine whether mixing synthetic vibration data with preprocessed vibration data helps improve the learning and generalisation of the classification model. Table 21 shows that the learning and validation losses decreased and the validation and testing accuracies increased. This shows that adding synthetic vibration data helps to improve the overall performance of the classification model.

Scenario 3 uses only synthetic vibration data for training and validation and only preprocessed vibration data for testing, rather than mixing the two vibration data as in Scenario 2. If the synthetic vibration data was similar enough to replace the preprocessed vibration data, it was expected that the classification model trained only on the synthetic vibration data would also show a high accuracy on the preprocessed vibration data in terms of generalisation to unseen data. However, as shown in Table 22, the testing accuracy did not improve as much as the validation accuracy and there was no improvement in the accuracy at all. From these results, it can be concluded that the two vibration data are not yet similar enough to be interchangeable.

Comparing the three scenarios, it is observed that when vibration data is insufficient, mixing synthetic vibration data with preprocessed vibration data as in Scenario 2 is effective in improving the classification model. Mixing too much synthetic vibration data will overly increase the dependence of the classification model on the hypothetical data generated using the generative model over the preprocessed vibration data obtained experimentally using the scaled model, so a moderate level of mixing of both vibration data is appropriate. In this study, 1,200 segments per label was chosen as appropriate given the slope of loss and accuracy and the practical stability of the learning process.

Figure 41 and Table 24 focus on the comparison between Scenario 1 and Scenario 2, with the exception of Scenario 3, using PCA and Euclidean distances between labels. Labels 2, 3 and 4 commonly correspond to wheel flats with the same depth of 2 mm, but due to their different geometries, the wheel flat-induced vibrations are also different and the centres of scatter for each label in PCA are relatively different. Therefore, these labels are located adjacent to each other in PCA with some overlap in terms of similarity.

In Figure 41 (a), for Scenario 1, the distance between Label 1 and Labels 2,3, and 4 in principal component 1 and principal component 2 is clear. In principal component 2 and principal component 3, Labels 2,3, and 4 also have their principal axes in different directions. In Figure 41 (b), the relationship between labels in the PCA plots is clearer in Scenario 2 because the synthetic vibration data with 1,200 segments per label was mixed with the preprocessed vibration data as described earlier. The Euclidean distance, which is the distance between labels in Table 24, also has a higher value in Scenario 2 than in Scenario 1. Both scenarios show quantitatively that Label 2, corresponding to 'Wheel Flat Type 1 - Simple Cut' has a clearer difference for Label 1, corresponding to 'Non-defect', compared to the other two labels.

From these observations, it is evident that Scenario 2, which uses a combination of the preprocessed and synthetic vibration data, provides a significant advantage in terms of classification performance over Scenario 1, which uses only the preprocessed vibration data. The improved distinction in PCA and the larger Euclidean distance in Scenario 2 indicate that the synthetic vibration data helps the classification model learn more distinct and clear features of each label, which helps improve the overall loss and accuracy and helps it generalise to unseen data.

5.4.3   Effects of Wheel Flat Modelling Method

From the PCA plots in Figure 41 and the Euclidean distance in Table 24, the three different wheel flat modelling methods (Labels 2, 3, and 4) can be analysed in comparison to the non-defect wheel model (Label 1) in terms of the scatter of each label. In both scenarios 1 and 2, Label 2 (Wheel Flat Type 1 - Simple Cut) has commonly a greater Euclidean distance compared to the other two labels. This means that physically, Label 2 produces more distinct wheel-flat-induced vibrations than the other two labels. The larger Euclidean distance indicates that the features of the wheel flats modelled by the simple cut method are more easily distinguishable from the features of the non-defect model, allowing for a more accurate classification by the deep learning model. Therefore, label 2 (Wheel Flat Type 1 - Simple Cut) is found to be the most effective in terms of disambiguation of the defect modelling method compared to the non-defect cases. It produces the most distinct differences in the feature space, facilitating the classification task and potentially increasing the classification accuracy for wheel flat detection.

Quantity of Preprocessed Vibration Data per Label	Training Loss	Validation Loss	Validation Accuracy	Testing Accuracy
200	0.169 ± 0.054	0.550 ± 0.101	0.806 ± 0.037	0.807 ± 0.034

Table 20: **Statistical Classification Performance for Scenario 1 as Baseline**

Quantity of Synthetic Vibration Data per Label	Training Loss	Validation Loss	Validation Accuracy	Testing Accuracy
200	0.065 ± 0.025	0.315 ± 0.071	0.911 ± 0.020	0.910 ± 0.021
400	0.047 ± 0.016	0.200 ± 0.052	0.943 ± 0.014	0.942 ± 0.014
600	0.037 ± 0.011	0.131 ± 0.041	0.967 ± 0.009	0.968 ± 0.009
800	0.034 ± 0.010	0.115 ± 0.035	0.969 ± 0.008	0.969 ± 0.009
1000	0.027 ± 0.008	0.075 ± 0.028	0.983 ± 0.006	0.983 ± 0.006
1200	0.028 ± 0.008	0.077 ± 0.024	0.980 ± 0.006	0.979 ± 0.006
1400	0.024 ± 0.008	0.064 ± 0.022	0.983 ± 0.005	0.984 ± 0.005
1600	0.023 ± 0.007	0.055 ± 0.019	0.985 ± 0.004	0.986 ± 0.004
1800	0.019 ± 0.006	0.040 ± 0.017	0.991 ± 0.003	0.991 ± 0.003
2000	0.021 ± 0.006	0.045 ± 0.016	0.988 ± 0.003	0.989 ± 0.004

Table 21: **Statistical Classification Performance for Scenario 2 by Quantity of Synthetic Vibration Data**

Quantity of Synthetic Vibration Data per Label	Training Loss	Validation Loss	Validation Accuracy	Testing Accuracy
200	0.019 ± 0.015	0.174 ± 0.094	0.950 ± 0.026	0.806 ± 0.057
400	0.017 ± 0.012	0.082 ± 0.054	0.980 ± 0.012	0.812 ± 0.068
600	0.018 ± 0.013	0.040 ± 0.032	0.991 ± 0.007	0.807 ± 0.073
800	0.017 ± 0.013	0.021 ± 0.024	0.995 ± 0.006	0.805 ± 0.076
1000	0.015 ± 0.011	0.010 ± 0.014	0.998 ± 0.003	0.803 ± 0.080
1200	0.013 ± 0.009	0.005 ± 0.009	0.999 ± 0.002	0.799 ± 0.079
1400	0.012 ± 0.009	0.003 ± 0.007	0.999 ± 0.002	0.803 ± 0.002
1600	0.011 ± 0.008	0.002 ± 0.008	1.000 ± 0.002	0.803 ± 0.085
1800	0.010 ± 0.008	0.001 ± 0.004	1.000 ± 0.001	0.797 ± 0.085
2000	0.010 ± 0.007	0.001 ± 0.003	1.000 ± 0.001	0.800 ± 0.084

Table 22: **Statistical Classification Performance for Scenario 3 by Quantity of Synthetic Vibration Data**

	Predicted Label 1	Predicted Label 2	Predicted Label 3	Predicted Label 4
True Label 1	0.990 ± 0.019	0.000 ± 0.000	0.006 ± 0.016	0.003 ± 0.011
True Label 2	0.000 ± 0.004	0.719 ± 0.107	0.149 ± 0.077	0.131 ± 0.080
True Label 3	0.040 ± 0.040	0.155 ± 0.081	0.734 ± 0.096	0.071 ± 0.052
True Label 4	0.000 ± 0.004	0.135 ± 0.081	0.077 ± 0.056	0.788 ± 0.097

(a) Scenario 1

	Predicted Label 1	Predicted Label 2	Predicted Label 3	Predicted Label 4
True Label 1	0.999 ± 0.002	0.000 ± 0.000	0.001 ± 0.002	0.000 ± 0.001
True Label 2	0.000 ± 0.001	0.992 ± 0.013	0.004 ± 0.009	0.004 ± 0.007
True Label 3	0.004 ± 0.004	0.003 ± 0.007	0.991 ± 0.010	0.001 ± 0.003
True Label 4	0.000 ± 0.001	0.004 ± 0.007	0.001 ± 0.003	0.994 ± 0.009

(a) Scenario 2

	Predicted Label 1	Predicted Label 2	Predicted Label 3	Predicted Label 4
True Label 1	0.981 ± 0.034	0.003 ± 0.015	0.011 ± 0.023	0.005 ± 0.019
True Label 2	0.003 ± 0.004	0.728 ± 0.339	0.153 ± 0.223	0.116 ± 0.164
True Label 3	0.055 ± 0.038	0.173 ± 0.246	0.716 ± 0.324	0.055 ± 0.089
True Label 4	0.004 ± 0.010	0.156 ± 0.220	0.055 ± 0.099	0.786 ± 0.266

(a) Scenario 3

Table 23:   **Statistical Confusion Matrices by Scenario**

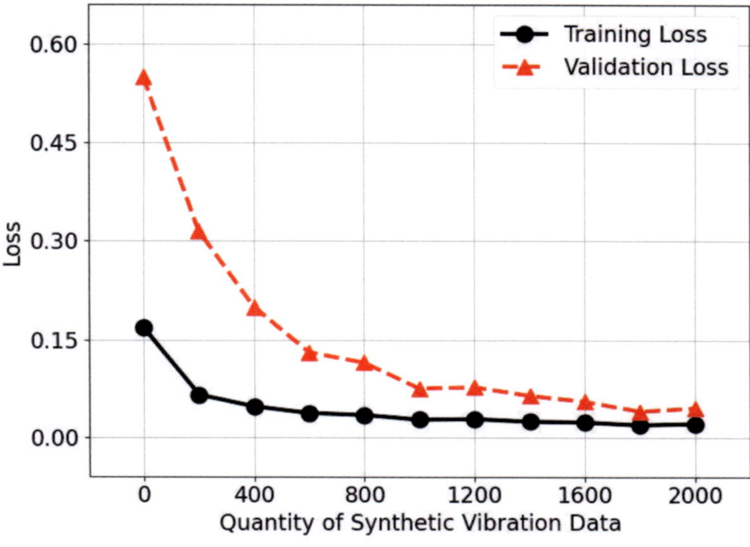

(a) Loss

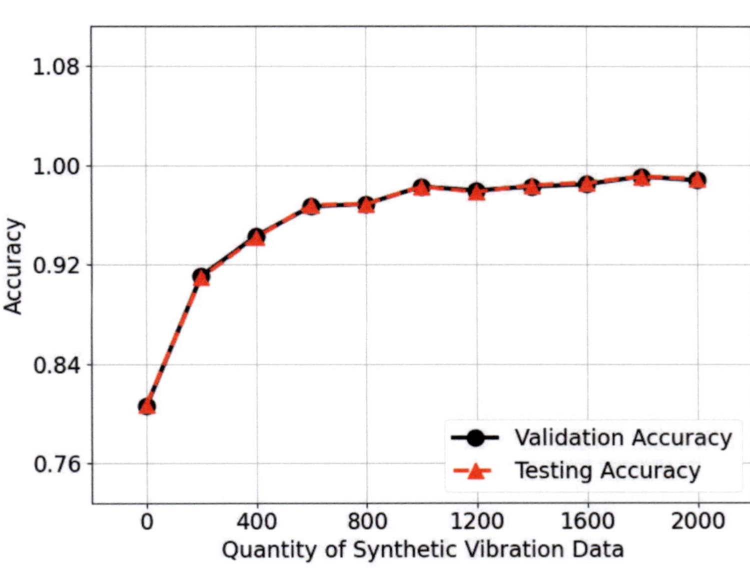

(b) Accuracy

Figure 35: **Effects of the Quantity of Synthetic Vibration Data on Training Loss and Accuracy for Scenario 2**

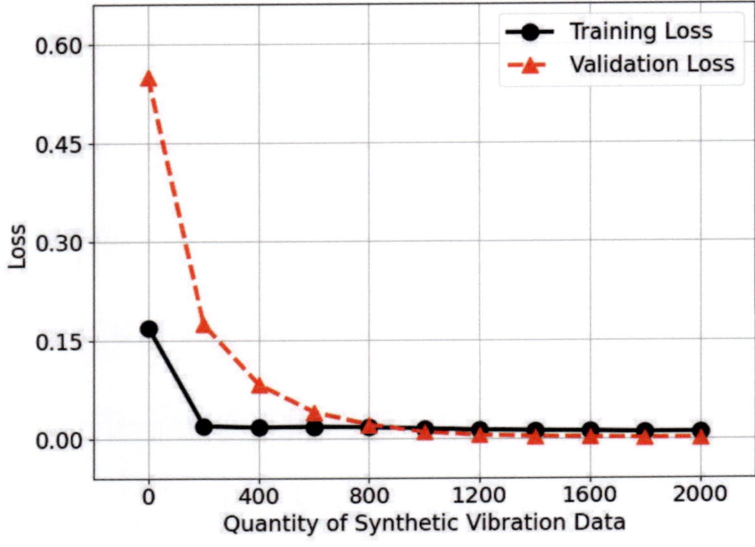

(a) Loss

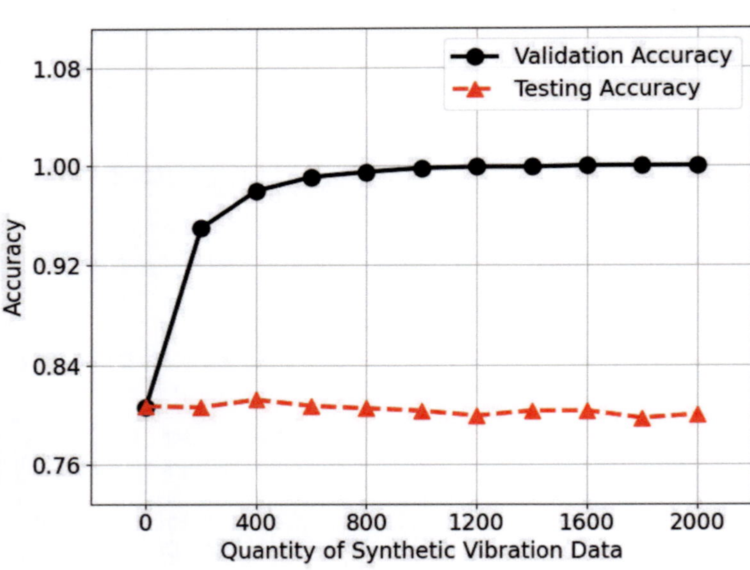

(b) Accuracy

Figure 36: **Effects of the Quantity of Synthetic Vibration Data on Training Loss and Accuracy for Scenario 3**

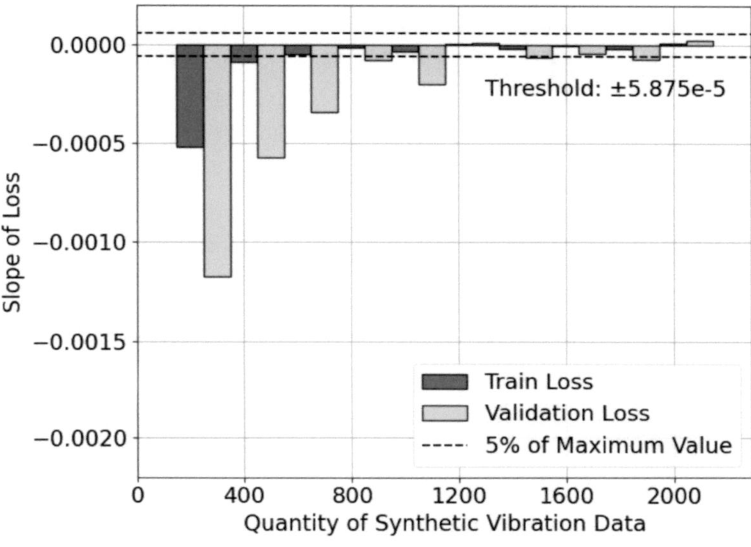

(a) Loss

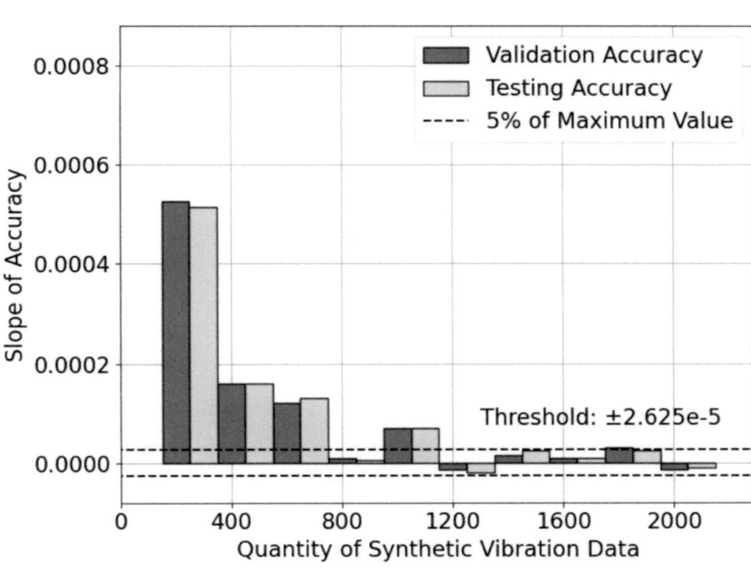

(b) Accuracy

Figure 37:   **Effects of the Quantity of Synthetic Vibration Data on Slope of Training Loss and Accuracy for Scenario 2**

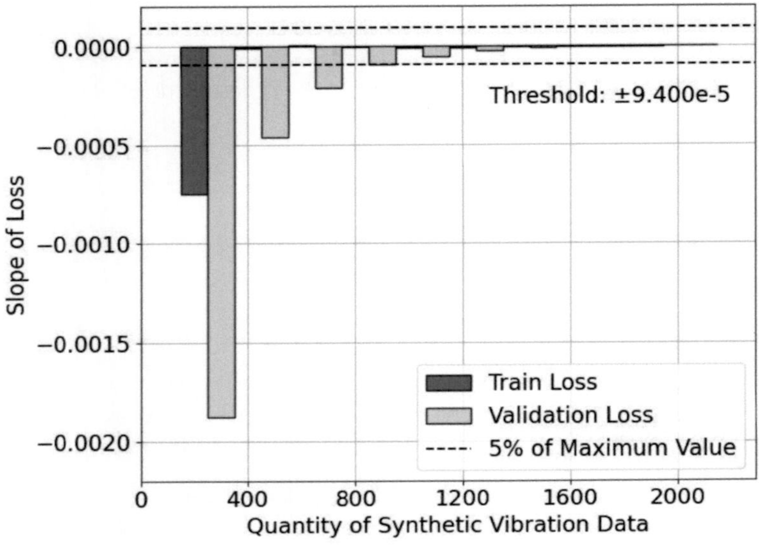

(a) Loss

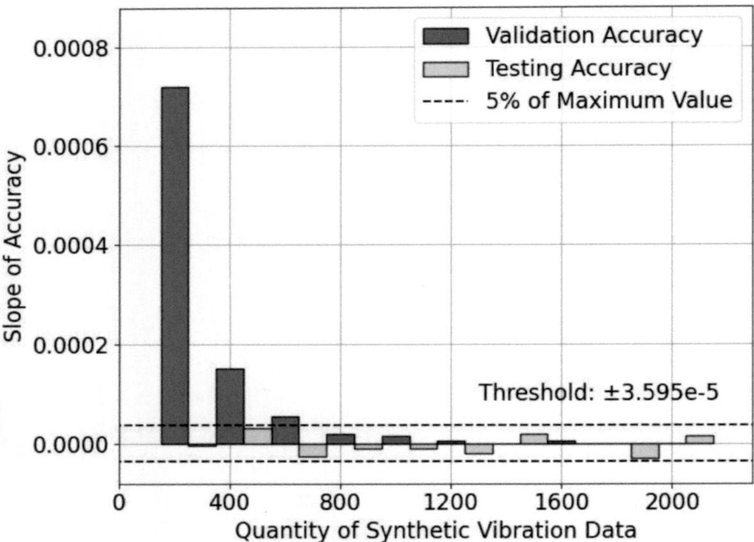

(b) Accuracy

Figure 38: **Effects of the Quantity of Synthetic Vibration Data on Slope of Training Loss and Accuracy for Scenario 3**

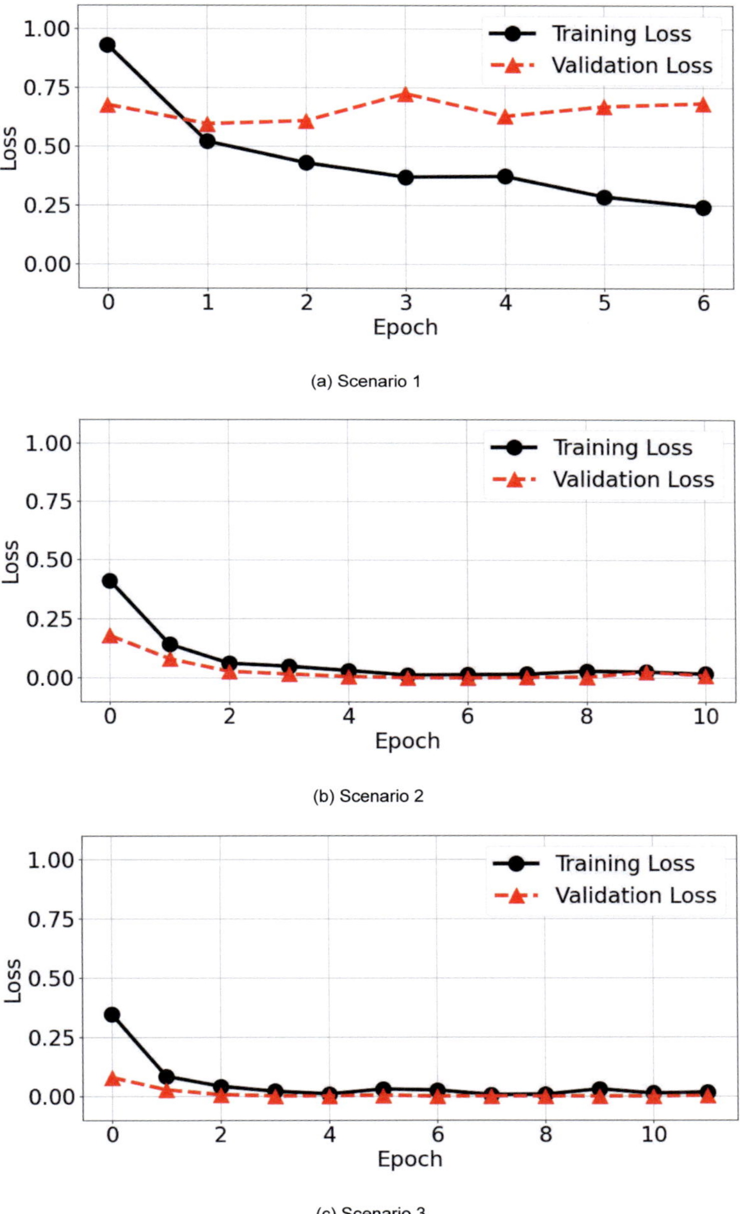

(a) Scenario 1

(b) Scenario 2

(c) Scenario 3

Figure 39: **Comparison of Training and Validation Loss for Best Hyperparameters on Additional 1,200 Segments per Label by Scenario**

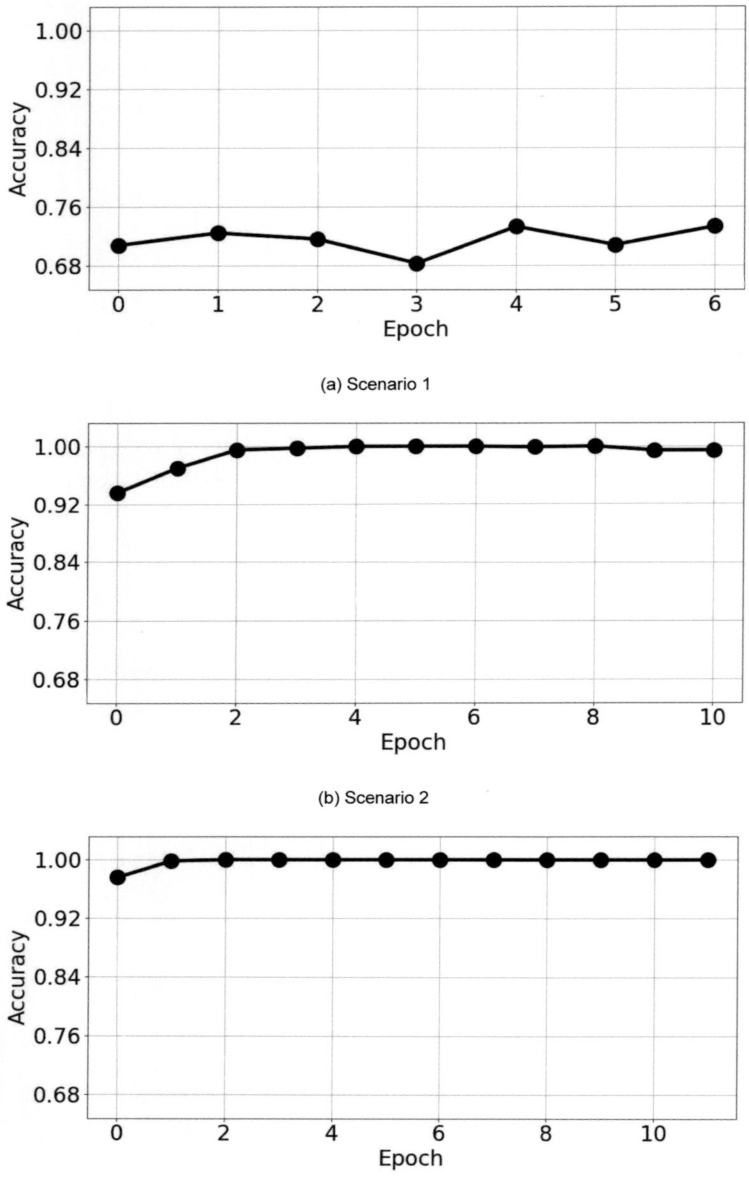

(a) Scenario 1

(b) Scenario 2

(c) Scenario 3

Figure 40: **Comparison of Validation Accuracy for Best Hyperparameters on Additional 1200 Synthetic Segments per Label by Scenario**

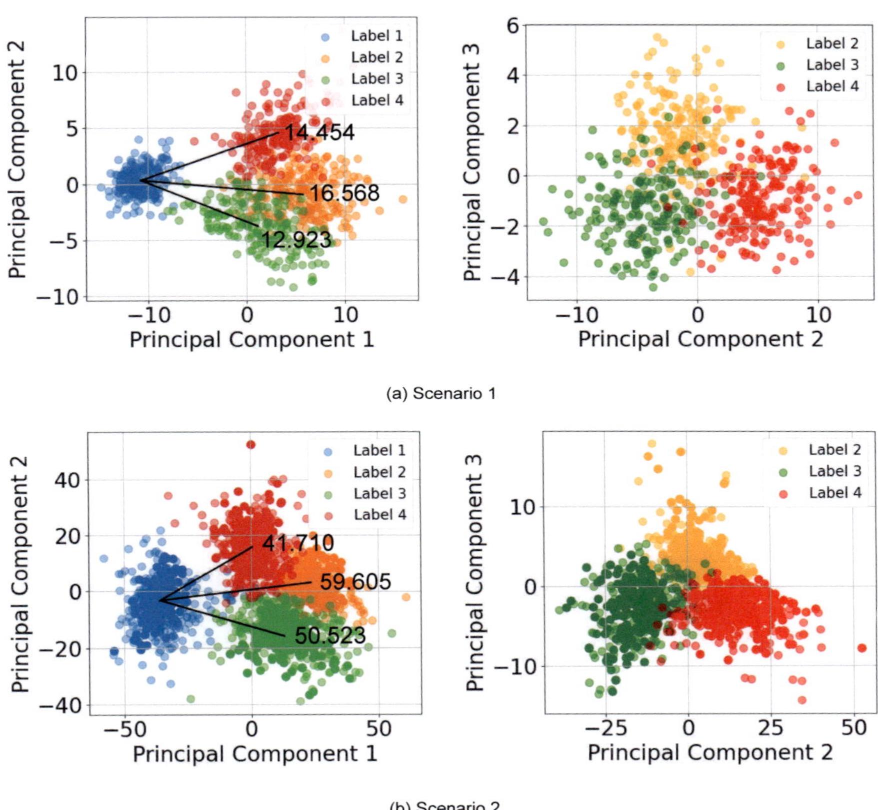

(a) Scenario 1

(b) Scenario 2

Figure 41: **Comparison of Principal Component Analysis before and after adding Synthetic Vibration Data**

Modelling Method	Label 2 Wheel Flat Type 1 - Simple Cut	Label 3 Wheel Flat Type 2 - Boolean Operation	Label 4 Wheel Flat Type 3 - FEM Simulation
Scenario 1	16.568	12.923	14.454
Scenario 2	59.605	50.523	41.710

Table 24: **Comparison of Euclidean Distance between Label 1 and the Rest of the Labels by Scenario**

6 Conclusion and Future Works

6.1 Summary and Conclusion

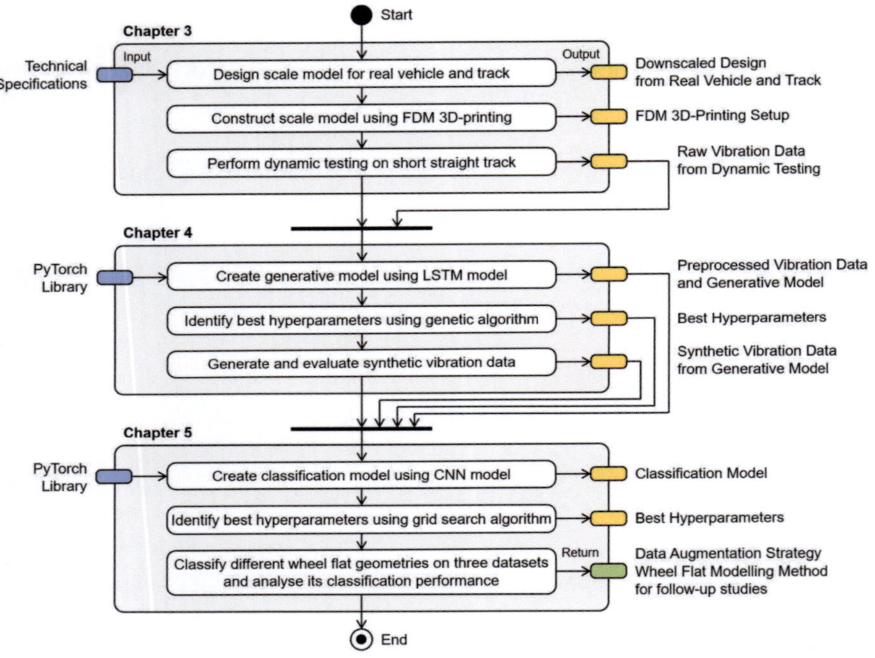

Figure 42: **Schematic Overview of Dissertation for Experimental Generation, Data Augmentation, and Classification of Wheel Flat-induced Vibration Data in Laboratory Environment**

This study was motivated by the difficulties in obtaining high-quality data from field experiments with real vehicles and tracks during the preliminary research stage, which aimed to develop data-driven detection algorithms for wheel flats using vibration data. While MBD simulations have been widely utilised as a numerical alternative in other studies, the railway scale model, constructed using FDM 3D-printing in the laboratory, was used to physically generate the vibration data due to wheel flats with different geometries. However, like field experiments, this approach also requires significant time and effort. To alleviate this, the vibration data was digitally augmented using the LSTM-based generative model, enhancing efficiency in data generation.

As others have mainly focused on 2D wheel flats for railway dynamics models, and there are few examples of 3D wheel flats, this study prioritised exploring the effects of different geometries of a single wheel flat on the vibration data rather than investigating the effects of different sizes. In the preparation process for creating a defective wheel with a single wheel flat, it was difficult to determine which of the wheel flat geometries was most suitable based on their geometries alone. Despite their different geometries, they share the same vibration mechanism and produce similar vibration data, and there is ambiguity in the classification process. This study explored how the vibration data could be augmented and combined with the preprocessed vibration data to obtain high-quality data, thereby improving the classification performance.

In this context, this dissertation focused on the implementation, augmentation, and classification of the vibration data obtained from the axle box accelerometer, which is positioned adjacent to the wheel flats on the 3D-printed railway scale model. The research activities and results of each subtopic are described in detail in Chapters 3, 4, and 5. Figure 42 illustrates the workflow of these research activities and outlines the inputs and outputs essential to the progress of this study.

Chapter 3 begins by choosing the law of similarity that maintains the same scale in time and frequency, but applies a different downscaling ratio between force and weight because gravity is the same before and after downscaling. The dimensions of the real metro vehicle and ballastless track were then downscaled to a 1/10 scale. There were inevitable design simplifications and compromises due to materials, manufacturing methods, budget, etc. In the design of the railway scale model, it was emphasized to consider the flexibility of the wheelset and the different modelling methods for the wheel flats. This is because the abnormal wheel-rail contact forces, caused by the wheel flats, are amplified by the structural bending modes of the flexible wheelset. As a result, the abnormal vibration energy, induced by the wheel flats, is transmitted through the wheelset and the bogie frame, and ultimately measured as abnormal acceleration by the axle box accelerometers.

During the construction process, almost all structural parts were printed in the laboratory using FDM 3D-printers and PLA filaments. Although the mechanical properties of 3D-printed parts are relatively poor and anisotropic compared to metal parts made by traditional machining methods, which have been mainly used in other studies. The fact that design mistakes can be easily fixed, and new ideas can be quickly incorporated during the construction process in the laboratory, without the need for outsourcing, is a significant advantage in terms of project efficiency and flexibility.

Lastly, the dynamic testing on the short straight track experimentally confirmed that the first bending mode of the flexible wheelset dominates the abnormal energy in the raw vibration data measured at the axle box. It was also confirmed that the modelling method of the wheel flats affects the signal shape in this data. Additionally, it was found that the 3D-printed parts are sufficiently robust to withstand the loads caused by track irregularities and wheel flats during operation on the track.

Chapter 4 described the process of creating the LSTM-based generative model to generate the synthetic vibration data based on the preprocessed vibration data for the four labels obtained from the dynamic testing. In the preprocessing, the raw vibration data was segmented, and then noisy segments were filtered out. In the training process, the hyperparameters were optimised using GA. The choice of hyperparameters was found to be critical to the quality of the synthetic vibration data. Then, the synthetic vibration data was then compared with the preprocessed vibration data and found to be statistically similar in terms of PSD and MAC. As a result, the LSTM-based generative model was found to be effective in generating the synthetic vibration data that was similar, but not identical, to the preprocessed vibration data.

Chapter 5 concludes this study by identifying the occurrence and type of wheel flats using the CNN-based classification model, which is based on the preprocessed and synthetic vibration data obtained in the previous two chapters. There are two main contributions. First, like Scenario 2, combining the synthetic vibration data with the preprocessed vibration data in the proper proportions was effective in improving high training loss and low testing accuracy, and in avoiding underfitting and overfitting during training process due to lack of vibration data. The required amount of synthetic vibration data seems to be 1,200 synthetic segments per label in terms of stability and results of the training process. When training and validating with only the synthetic vibration data and then testing with only the preprocessed vibration data, like Scenario 3, the training loss progressively decreased as the amount of the synthetic vibration data increased, but the test accuracy did not change at all. This means that the two types of vibration data are not fully interchangeable enough to replace each other.

Secondly, the appropriate modelling method for the wheel flats can vary depending on several: quick and easy fabrication, consistent vibration response, and similarity to the actual geometry. Initially, it was expected that modelling the geometry of the wheel flat as realistically as possible, as in 'Wheel Flat Type 3 - FEM Simulation', would help to obtain consistent vibration responses. However, the actual transient vibration responses caused by wheel flats were highly inconsistent. Furthermore, each defect condition would require a time-consuming FEM simulation for fabrication. Next, 'Wheel Flat Type 2 - Boolean Operation' also shows inconsistent vibration responses. Although this model uses Boolean operations to speed up fabrication, misalignments on the rail can lead to unintended lateral forces at the wheel-rail contact, due to overlaps that do not actually exist.

In contrast, 'Wheel Flat Type 1 - Simple Cut' shows consistent vibration responses. Its simpler modelling approach reduces fabrication time significantly. Although the wheel flats are somewhat larger than the actual geometry, this is expected to be compensated by a relatively higher material damping ratio than steel. While collecting vibration data for conceptual defect detection studies in a laboratory environment, the quick fabrication and consistent vibration responses are more important than precise geometric similarity. From these observations, it was concluded that 'Wheel Flat Type 1 - Simple Cut' was identified as the most suitable modelling method for implementing practical defect detection algorithms in follow-up studies.

Quickly conceptualising ideas for structural defect detection in a lab environment and refining them based on quick results, feedback and discussion is essential to ensure meaningful results and to save both time and costs in field experiments with real railway vehicles and tracks. This study tried to overcome the difficulties of obtaining high-quality data in this research process by developing the 3D-printed railway scale model and the LSTM-based generative model. In terms of the 3V (volume, variety, and velocity), the railway scale model is effective in improving the volume and variety, while the generative model is effective in improving the volume and velocity. The effects of the bending modes of the wheelset and the geometry of the wheel flats were also reflected in the vibration data as intended during the design process. In conclusion, these achievements are expected to help make the data preparation process easier than before.

6.2   Future Works

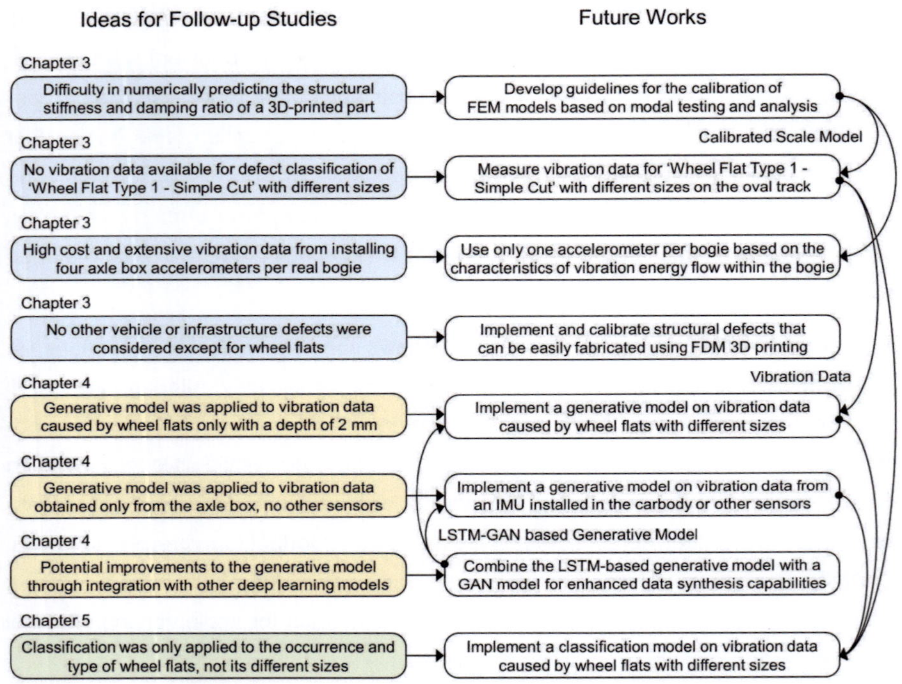

Figure 43:   Ideas for Future Works

As this study focuses on the detection of the wheel flats using the vibration data, the 3D-printed railway scale model was only used to physically generate abnormal vibration data at the location of the axle box caused by a wheel flat of a certain size. This can also be used for follow-up studies to deepen the existing study or for new studies to implement and detect structural defects in railway vehicles or infrastructure. In this context, ideas for follow-up studies are listed in Figure 43. Possible future works are suggested as follow:

1. Since FDM 3D-printing technology is based on the principle of layer-by-layer deposition, it has inherently anisotropic mechanical properties, and is usually

filled with an infill structure to reduce printing time and save material. The filament material, 3D-printing settings, and infill structure pattern have a significant influence on the structural strength, stiffness, and damping ratio of a 3D-printed part. Among these, the stiffness and damping ratio have a strong influence on vibration data. Building a finite element model of the infill structure as it appears requires significant time and cost during the modelling and computation process. In addition, the material properties provided by filament manufacturers are not always available. This makes it difficult to make numerical predictions using traditional FEM-based approaches. Alternatively, guidelines for accurate and efficient FEM simulation based on experimental modal testing and analysis are necessary. This will help to generate vibration data from the 3D-printed railway scale model in the intended shape and understand the structural vibration characteristics of the bogie structure.

2. In the process of determining the appropriate geometry of the wheel flats, only a depth of 2 mm was considered. This value was obtained from the downscaled value of the wheel flats with a critical size that clearly requires maintenance according to evaluation standards or criteria. The value was then doubled as large due to the relatively high damping ratio of PLA filaments compared to steel. In order to implement a deep learning-based classification model for the wheel flats with different sizes, within the range given in evaluation standards or criteria, vibration data needs be prepared experimentally using 'Wheel Flat Type 1 - Simple Cut' with different sizes.

3. Installing a large number of accelerometers and measurement equipment throughout an entire railway vehicle, with four axle boxes per bogie, requires high cost and effort. It is difficult to continuously process the large amount of vibration data obtained from these accelerometers. Alternatively, it is necessary to monitor the four axle boxes with one accelerometer or other type of single sensor, based on the understanding of the vibration energy flow within the bogie structure. This requires information from modal analysis and frequency response analysis, either experimentally or numerically.

4. In order to expand the application range of the 3D-printed railway scale model, it is necessary to investigate structural defects in railway vehicles or infrastructure, which can be easily implemented using FDM 3D-printing technology.

5. The LSTM-based generative model was implemented using the vibration data from the wheel flats with a depth of 2 mm. For broader applicability, this generative model needs to be trained on new vibration data from the wheel flats with different sizes.

6. The LSTM-based generative model was implemented using the vibration data measured using accelerometers at the location of the axle box. This generative model could be implemented using new vibration data measured from an IMU located at the centre of the carbody or other sensors.

7. In the process of implementing LSTM-based generative models, the training process is important because the optimisation of the hyperparameters significantly affects the quality of the synthetic vibration data. To achieve better results, combining the advantages of different deep learning models used as a hybrid generative model could be a good way. For example, GANs, which are popular among generative models, can be combined with the LSTM network. LSTM-GAN based generative models can be used for better training of new vibration data as mentioned above.

8. In this study, the CNN-based classification model was applied to the vibration data from the wheel flats with a depth of 2 mm in terms of evaluating the data quality during the generation and augmentation of vibration data based on the 3D-printed railway scale model. Through this process, the appropriate modelling method 'Wheel Flat Type 1 - Simple Cut' of the wheel flat geometry was determined. In the above-mentioned future works, the enhanced diversity of the vibration data and its augmentation data can be evaluated on a CNN or other deep learning model-based classification model.

Glossary

Axle Box
A structural part that is positioned at both ends of a wheelset and supports the bogie and carbody through primary and secondary suspensions. IMUs were installed on the side of the axle box to measure the axle box acceleration.

Bogie
A structural frame that is positioned beneath the carbody, containing traction motors and wheelsets at the front and rear. Due to limited space, the scale model was designed for independent operation at the bogie level, not the carbody.

Butterworth Filter
A signal filter with a flat passband and minimal response ripple, which was used as a high-pass filter to remove low-frequency components from raw vibration data.

FDM 3D-Printer
An additive manufacturing equipment using the Fused Deposition Modelling (also known as Fused Filament Manufacturing) method, which melts thermoplastic material to build objects layer by layer. This was used to print most structural parts of the scale model.

Genetic Algorithm
An optimisation model inspired by natural selection during the evolutionally process, which was applied to optimise hyperparameters of LSTM-based generative models.

Grid Search Algorithm
An optimisation model that explores the entire parameter space, which was applied to optimise hyperparameters of CNN-based classification models.

Gyroid Infill Structure
An infill structure for isotropic mechanical properties and rapid printing, though it exhibits higher vibration during printing compared to other infill structures. This was uniformly applied across all 3D-printed parts.

High-quality Data
Vibration data in data-driven approaches adequately satisfies the 3V's of big data: volume, variety, and velocity.

Law of Similarity	A law defining relationships between physical variables before and after downscaling. Several laws exist depending on which variables are considered significant. The law of using the same scale for time and frequency was applied.
Modal Properties	Variables that characterise structural resonances in terms of resonance frequency, mode shape, and damping ratio. Since the bending modes of the wheelset contribute significantly to the overall vibration, their modal properties are considered important.
Raw, Preprocessed, and Synthetic Vibration Data	The terms 'raw', 'preprocessed', and 'synthetic' were used progressively as prefixes to denote the data stages of the vibration data during dynamic testing, preprocessing, and augmentation.
Shannon Entropy	An indicator for quantifying the amount of information, which was applied for terminating further processes when there is no diversity of genetic information among individuals in an evolutionary process using GA.
Signal-to-Noise Ratio	The relative energy ratio between a transient vibration component due to wheel flats and noise (remaining components) in the raw vibration data.
Track Irregularities	Unevenness or deviation from track geometry, typically characterised by its PSD, results in a force exerted on a vehicle at the wheel-rail contact, which is always present regardless of whether a defect is present or not.
Wheelset	It consists of two wheels and an axle. Its bending modes are closely related to the vertical vibration response of a vehicle. The first and second bending mode of the 3D-printed wheelset was considered important.

Bibliography

Bahamon-Blanco, S., S. Rapp, C. Rupp, J. Liu, and U. Martin. "Recognition of track defects through measured acceleration - Part 1 & Part 2." *7th International Conference of Euro Asia Civil Engineering Forum.* Stuttgart, Germany, 2019. 012121.

Bahamon-Blanco, Sebastian, Jing Liu, and Ullrich Martin. "Convolutional Neural Network for the Early Determination of Local Instabilities." *Proceedings of IRSA 2021 - 3rd International Railway Symposium.* Aachen, Germany, 2021. 70-81.

Bahamon-Blanco, Sebastian, Sebastian Rapp, Yi Zhang, Jing Liu, and Ullrich Martin. "Recognition of Track Defects through Measured Acceleration Using A Recurrent Neural Network." *International Journal of Computational Methods and Experimental Measurements,* 2020: 270-280.

Bosso, Nicola, Antonio Gugliotta, and Nicolò Zampieri. "Wheel flat detection algorithm for onboard diagnostic." *Measurement,* 2018: 193-202.

Brizuela, J., C. Fritsch, and A. Ibáñez. "Railway wheel-flat detection and measurement by ultrasound." *Transportation Research Part C: Emerging Technologies,* 2011: 975-984.

Bruni, Stefano, Jordi Vinolas, Mats Berg, Oldrich Polach, and Sebastian Stichel. "Modelling of suspension components in a rail vehicle dynamics context." *International Journal of Vehicle Mechanics and Mobility,* 2011: 1021-1072.

Bu, Seok-Jun, Hyung-Jun Moon, and Sung-Bae Cho. "Adversarial Signal Augmentation for CNN-LSTM to Classify Impact Noise in Automobiles." *IEEE International Conference on Big Data and Smart Computing.* Jeju Island, Korea, 2021.

Cai, Wubin, Maoru Chi, Gongquan Tao, Xingwen Wu, and Zefeng Wen. "Experimental and Numerical Investigation into Formation of Metro Wheel Polygonalization." *Shock and Vibration,* 2019: Article ID 1538273.

Chamorro, Rosario, Javier F. Aceituno, Pedro Urda, Enrique del Pozo, and José L. Escalona. "Design and manufacture of a scaled railway track with mechanically variable geometry." *Scientific Reports,* 2022: 8665.

Chen, Yanxiang, et al. "Wheel fault diagnosis model based on multichannel attention and supervised contrastive learning." *Advances in Mechanical Engineering*, 2021.

Guo, Qingwen, Yibin Li, Yanjun Liu, Shengyao Gao, and Yan Song. "Data Augmentation for Intelligent Mechanical Fault Diagnosis Based on Local Shared Multiple-Generator GAN." *IEEE Sensors Journal*, 2022: 9598-9609.

Hung, C., et al. "Study on detection of the early signs of derailment for railway vehicles." *International Journal of Vehicle Mechanics and Mobility*, 2010: 451-466.

Jaschinski, A., H. Chollet, S., Iwnicki, A., Wickens, and J. Würzen. "The Application of Roller Rigs to Railway Vehicle Dynamics." *International Journal of Vehicle Mechanics and Mobility*, 1999: 345-392.

Jia, S., and M. Dhanasekar. "Detection of Rail Wheel Flats using Wavelet Approaches." *Structural Health Monitoring*, 2007: 1475-9217.

Kim, Min-Soo, Joon-Hyuk Park, and Won-Hee You. "Construction of active steering control system for the curving performance analysis of the scaled railway vehicle." *Proceedings of the 7th conference on Circuits, systems, electronics, control and signal processing.* Wisconsin, United States, 2008. 223-227.

Krizhevsky, Alex, Ilya Sutskever, and Geoffrey E. Hinton. "ImageNet Classification with Deep Convolutional Neural Networks." *Advances in neural information processing systems*, 2012.

Krummenacher, Gabriel, Cheng Soon Ong, Stefan Koller, Seijin Kobayashi, and Joachim M. Buhmann. "Wheel Defect Detection With Machine Learning." *IEEE Transactions on Intelligent Transportation Systems*, 2017: 1176 - 1187.

Kurzeck, Bernhard, and Luciano Valente. "A novel mechatronic running gear: concept, simulation and scaled roller rig testing." *9th World Congress on Railway Research.* Lille. France, 2011.

Laney, D. *3-D Data Management: Controlling Data Volume, Velocity and Variety.* Gartner, Inc, 2001.

Li, Yifan, Jianxin Liu, and Yan Wang. "Railway Wheel Flat Detection Based on Improved Empirical Mode Decomposition." *Shock and Vibration*, 2016: 4879283 .

Liang, B., S. D. Iwnicki, Y. Zhao, and D. Crosbee. "Railway wheel-flat and rail surface defect modelling and analysis by time–frequency techniques." *International Journal of Vehicle Mechanics and Mobility*, 2013: 1403-1421.

Liu, Shaowei, Hongkai Jiang, Zhenghong Wu, and Xingqiu Li. "Data synthesis using deep feature enhanced generative adversarial networks for rolling bearing imbalanced fault diagnosis." *Mechanical Systems and Signal Processing*, 2022: 108139.

Liu, Shaowei, Hongkai Jiang, Zhenghong Wu, and Xingqiu Li. "Rolling bearing fault diagnosis using variational autoencoding generative adversarial networks with deep regret analysis." *Measurement*, 2021: 108371.

Liu, Yunpeng, Hongkai Jiang, Renhe Yao, and Hongxuan Zhu. "Interpretable data-augmented adversarial variational autoencoder with sequential attention for imbalanced fault diagnosis." *Journal of Manufacturing Systems*, 2023: 342-359.

Lu, Zheng-Gang, and Zhe, Qi Huang, Xiao-Chao Wang Yang. "Robust active guidance control using the μ-synthesis method for a tramcar with independently rotating wheelsets." *Proceedings of the Institution of Mechanical Engineers, Part F: Journal of Rail and Rapid Transit*, 2019: 33-48.

Ma, Liang, Yu Ding, Zili Wang, Chao Wang, Jian Ma, and Chen Lu. "An interpretable data augmentation scheme for machine fault diagnosis based on a sparsity-constrained generative adversarial network." *Expert Systems with Applications*, 2021: 115234.

Ma, Xirui, Yizhou Lin, Zhenhua Nie, and Hongwei Ma. "Structural damage identification based on unsupervised feature-extraction via Variational Auto-encoder." *Measurement*, 2020: 107811.

Michitsuji, Y., and Y. Suda. "Running performance of power-steering railway bogie with independently rotating wheels." *International Journal of Vehicle Mechanics and Mobility*, 2006: 71-82.

Mohammadi, Mohammadreza, Araliya Mosleh, Cecilia Vale, Diogo Ribeiro, and Pedro Montenegro. "An Unsupervised Learning Approach for Wayside Train Wheel Flat Detection." *Sensors*, 2023: 1910.

Mosleh, Araliya, Andreia Meixedo, Diogo Ribeiro, Pedro Montenegro, and Rui Calçada. "Early wheel flat detection: an automatic data-driven wavelet-based approach for railways." *International Journal of Vehicle Mechanics and Mobility*, 2023: 1644-1673.

Mosleh, Araliya, Pedro Aires Montenegro, Pedro Alves Costa, and Rui Calçada. "Railway Vehicle Wheel Flat Detection with Multiple Records Using Spectral Kurtosis Analysis." *Applied Sciences*, 2021: 4002.

Mosleh, Araliya, Pedro Montenegro, Pedro Alves Costa, and Rui Calçada. "An approach for wheel flat detection of railway train wheels using envelope spectrum analysis." *Maintenance, Management, Life-Cycle Design and Performance*, 2020: 1710-1729.

Naeimi, Meysam, Zili Li, Roumen H. Petrov, Jilt Sietsma, and Rolf Dollevoet. "Development of a New Downscale Setup for Wheel-Rail Contact Experiments under Impact Loading Conditions." *Experimental Techniques*, 2017: 1-17.

Naeimi, Meysam, Zili Li, Roumen H. Petrov, Jilt Sietsma, and Rolf Dollevoet. "Development of a New Downscale Setup for Wheel-Rail Contact Experiments under Impact Loading Conditions." *Experimental Techniques*, 2018: 1-17.

Ning, Jing, Yunzhi Ning, Jiangang Zhao, Yikun Yang, Yanping Li, and Chunjun Chen. "Small-Amplitude Hunting Motion Detection in High-Speed Trains Using Dcvae-Gan Under Extreme Data Imbalance." *Social Science Research Network*, 2022.

Oh, Yejun, Jae-Kwang Lee, Huai-Cong Liu, Sooyoung Cho, Ju Lee, and Ho-Joon Lee. "Hardware-in-the-Loop Simulation for Active Control of Tramcars With Independently Rotating Wheels." *IEEE Access*, 2019: 71252 - 71261.

Pieringer, A., W. Kropp, and J.C.O. Nielsen. "The influence of contact modelling on simulated wheel/rail interaction due to wheel flats." *Wear*, 2014: 273-281.

Rashid, Khandakar M., and Joseph Louis. "Times-series data augmentation and deep learning for construction equipment activity recognition." *Advanced Engineering Informatics*, 2019: 100944.

Sebastian Rapp, Ullrich Martin, Marius Strähle, Moritz Scheffbuch. "Track-vehicle scale model for evaluating local track defects detection methods." *Transportation Geotechnics*, 2019: 9-18.

Shah, Milind, Vinay Vakharia, Rakesh Chaudhari, Jay Vora, Danil Yu. Pimenov, and Khaled Giasin. "Tool wear prediction in face milling of stainless steel using singular generative adversarial network and LSTM deep learning models." *The International Journal of Advanced Manufacturing Technology*, 2022: 723-736.

Shi, Dachuan, Yunguang Ye, Marco Gillwald, and Markus Hecht. "Designing a lightweight 1D convolutional neural network with Bayesian optimization for wheel flat detection using carbody accelerations." *International Journal of Rail Transportation*, 2021: 311-341.

Shi, Dachuan, Yunguang Ye, Marco Gillwald, and Markus Hecht. "Robustness enhancement of machine fault diagnostic models for railway applications through data augmentation." *Mechanical Systems and Signal Processing*, 2022: 108217.

Shim, Jaeseok, Jeongseo Koo, Yongwoon Park, and Jaehoon Kim. "Anomaly Detection Method in Railway Using Signal Processing and Deep Learning." *Applied Sciences*, 2022: 12901.

Shin, Yu-Jeong, Won-Hee You, Hyun-Moo Hur, Joon-Hyuk Park, and Gyu-Seop Lee. "Improvement of Ride Quality of Railway Vehicle by Semiactive Secondary Suspension System on Roller Rig Using Magnetorheological Damper." *Advances in Mechanical Engineering*, 2014.

Singh, Rathore Maan, and S.P. Harsha. "Non-linear Vibration Response Analysis of Rolling Bearing for Data Augmentation and Characterization." *Journal of Vibration Engineering & Technologies*, 2023: 2109-2131.

Sresakoolchai, Jessada, and Sakdirat Kaewunruen. "Wheelflat detection and severity classification using deep learning techniques." *Insight - Non-Destructive Testing and Condition Monitoring*, 2021: 393-402.

Takikawa, M., and Y. Iriya. "Laboratory simulations with twin-disc machine on head check." *Wear*, 2008: 1300-1308.

Timo, Koenig, Cadau Luca, Wagner Fabian, and Kley Markus. "A generative adversarial network-based data augmentation approach with transient vibration data." *Procedia Computer Science*, 2023: 1340-1349.

Uzzal, R.U.A., A.K.W. Ahmed, and R.B. Bhat. "Modelling, validation and analysis of a three-dimensional railway vehicle–track system model with linear and nonlinear track properties in the presence of wheel flats." *International Journal of Vehicle Mechanics and Mobility*, 2013: 1695-1721.

Vuong, T.T., et al. "Investigation of a transitional wear model for wear and wear-type rail corrugation prediction." *Wear*, 2011: 287-298.

Wang, Shaohua, Lihua Tang, Yinling Dou, Zhaoyu Li, and Kean C. Aw. "Enhancement of Track Damage Identification by Data Fusion of Vibration-Based Image Representation." *Journal of Nondestructive Evaluation*, 2023.

Xie, Jiawei, Jinsong Huang, Cheng Zeng, Shui-Hua Jiang, and Nathan Podlich. "Systematic Literature Review on Data-Driven Models for Predictive Maintenance of Railway Track: Implications in Geotechnical Engineering." *Geosciences*, 2020: 425.

Ye, Yunguang, Caihong Huang, Jing Zeng, Yichang Zhou, and Fansong Li. "Shock detection of rotating machinery based on activated time-domain images and deep learning: An application to railway wheel flat detection." *Mechanical Systems and Signal Processing*, 2023: 109856.

Zhu, J. Y., D. J. Thompson, and C. J.C. Jones. "On the effect of unsupported sleepers on the dynamic behaviour of a railway track." *International Journal of Vehicle Mechanics and Mobility*, 2011: 1389-1408.

Zunsong, Ren. "An investigation on wheel/rail impact dynamics with a three-dimensional flat model." *International Journal of Vehicle Mechanics and Mobility*, 2019: 369-388.

List of Abbreviations

3V	Volume, Variety, and Velocity in the properties of big data
BLDC	Brushless Direct Current
C3D20	Second-order Hexahedral Element in Abaqus CAE
CA	Cement and Emulsified Asphalt
CNN	Convolutional Neural Network
CWT	Continuous Wavelet Transform
EMD	Empirical Mode Decomposition
FDM	Fused Deposition Modelling
FEM	Finite Element Method
FFT	Fast Fourier Transform
GA	Genetic Algorithm
GAN	Generative Adversarial Network
GPU	Graphics Processing Unit
HT	Hilbert Transform
I^2C	Inter-Integrated Circuit
IMU	Inertial Measurement Unit
LSTM	Long Short-Term Memory
MAC	Modal Assurance Criterion
MBD	Multi-body Dynamics
MSE	Mean Squared Error
NTP	Network Time Protocol
PCA	Principal Component Analysis
PLA	Polylactic Acid

PSD	Power Spectral Density
PWM	Pulse Width Modulation
ReLU	Rectified Linear Unit
ResNet	Residual Neural Network
RMS	Root Mean Square
RNN	Recurrent Neural Network
RTC	Real Time Clock
SNR	Signal-to-Noise Ratio
STFT	Short-time Fourier Transform
SVM	Support Vector Machine
TPU	Thermoplastic Polyurethane
VAE	Variational Autoencoder
WE	Wavelet Decomposition
WPT	Wavelet Packet Transform

List of Formulas

$$E(x) = \sum_{i=1}^{N} x_i^2$$

$E(x)$ is the energy of a segment, where N is the length of the segment, and the energy is calculated as the sum of the squares of the values x^i.

$$\alpha = \frac{E_{processed}(x_t)}{E_{synt\square etic}(y_t)}$$

α is the energy correction factor, which is the ratio of the energy of the preprocessed input x_t to the energy of the synthetic output y_t from a LSTM cell.

$$y_t' = \alpha \cdot y_t$$

The corrected synthetic output y_t' is calculated by the product of the energy correction factor α and the synthetic output y_t from a LSTM cell.

$$L_{avg}$$

L_{avg} is the average of the training losses from four LSTM cells.

$$fitness = \frac{1}{1 + L_{avg}}$$

$fitness$, defined as the reciprocal of $1+L_{avg}$, is the measure of an individual's suitability within GA, employed for its ability to provide positive, non-infinite values compared to using L_{avg} alone.

$$H(X_i) =$$
$$-\sum_{j=1}^{n} P(x_{ij}) \cdot log_2\left(P(x_{ij})\right)$$

The Shannon entropy $H(X_i)$ is the measure of the average amount of information produced by a stochastic source of data, quantifying the uncertainty involved in predicting the value of X_i. This is used in GA to quantify the diversity of genetic information within a population.

$$H_{total}(X)$$

$H_{total}(X)$ is the sum of the Shannon entropies from four LSTM cells. If this value is close to zero, it means that the genetic information within a population has converged on a particular combination of genetic information.

$$MAC(f) = \frac{|X(f) \cdot Y'(f)|^2}{|X(f)|^2 \cdot |Y'(f)|^2}$$

The Modal Assurance Criterion $MAC(f)$, is a statistical measure that quantifies the degree of correlation between two mode shapes in structural dynamics. In this study, $MAC(f)$ is used to measure the cosine similarity in the frequency domain between the preprocessed input $X(f)$ and the corrected output $Y'(f)$ from a LSTM cell.

$$MAC_{avg} = \frac{1}{N} \sum_{f=1}^{N} MAC(f)$$

MAC_{avg} is the average MAC calculated across a span of N distinct frequency points.

Appendix I: Downscaling Strategies by Research Objective

Related studies on the design and construction of scale models frequently have been mentioned the law of similarity to properly connect their scale models with reality in downscaling. Since there is no law that can connect perfectly for all cases, the law of similarity is defined based on different assumptions according to physical variables considered important. In general, three laws are frequently mentioned in related studies, and the relationship between physical variables is not always perfect after downscaling. Therefore, related studies have been appropriately selected according to their purpose. Figure 43 shows the summary for the three existing laws of similarity.

The explanations of each law are as follows:

The first law is defined for the same scale in time and frequency and it is widely used in related studies for vehicle dynamics on track or roller. The weak point arises because after downscaling, the force and vehicle weight have different downscale ratios and the gravity remains the same as a constant. As a result, the force by vehicle weight is less downscaled than other forces. Alternatively, vertical wires are often attached to axle boxes to reduce the influence of vehicle weight and to solve this conflict.

The second law is defined for the same scale in stress and it is used in related studies for wheel-rail contact and stress on track or roller. The relationship between physical variables after downscaling coincide with modal properties well, but velocity has still the same scale. Thus, too high a vehicle velocity is required for a study on vehicle dynamics. The test rig may be damaged due to too high stress during the test running.

The third law is defined for non-linear vehicle lateral dynamics (hunting) on roller and it is used in related studies for vehicle dynamics. In order for this law to be perfect, only the density among material properties must be downscaled. This conflict is solved by using a compromised value because it is difficult to decrease only the density in reality.

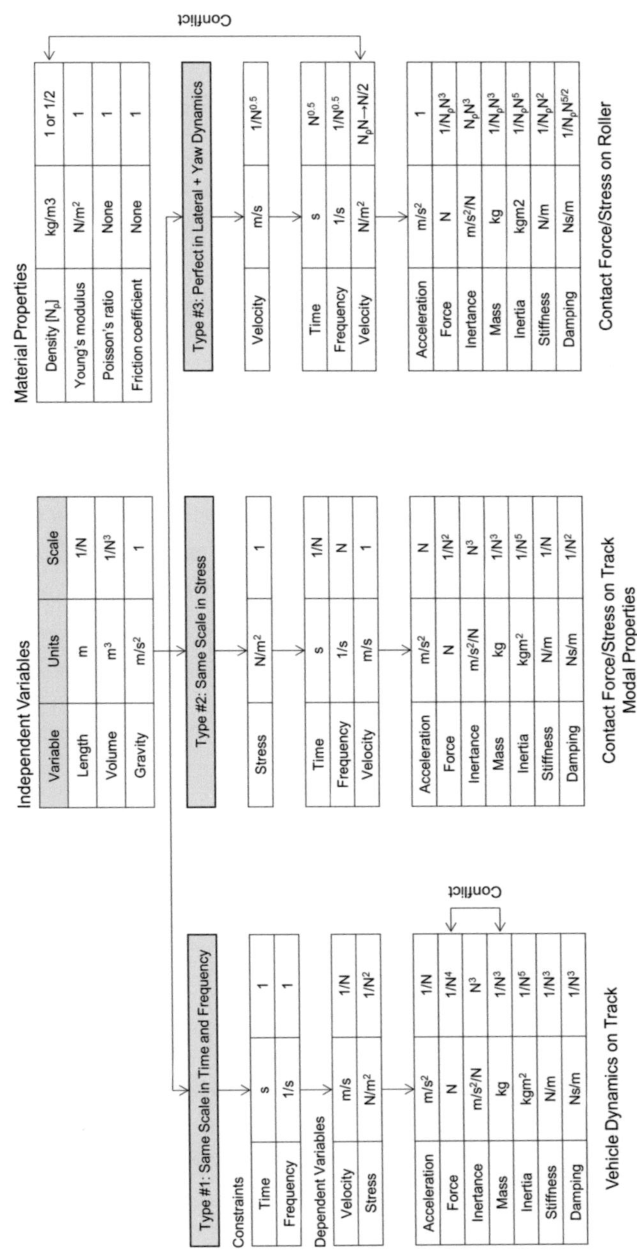

Figure 44: **Summary for Three Existing Laws of Similarity**

Appendix II: Fused Deposition Modelling (FDM) 3D-Printers

For the purpose of this study, Polylactic Acid (PLA) filaments were used because it is easier to handle than polyamide-based filaments containing short carbon fibres, which have high stiffness and strength but are expensive and difficult to print. Only PLA+ from eSUN was used, as the material properties of each filament manufacturer vary slightly. PLA+ does not require high nozzle temperatures during printing, is not sensitive to ambient temperature, and does not have a high-water absorption rate. Therefore, it was initially expected that a low-cost desktop FDM printer would be sufficient rather than an expensive industrial FDM printer.

The selection of a 3D-printer was based on motion mechanism, print resolution, speed, durability, consistency, serviceability, build volume, slicer software, and price. Initially, Creality's CR-10 Smart was used with slicer software UltiMaker Cura. Later, Bambu Lab P1P was mainly used with slicer software Prusa Slicer for faster and more accurate printing. However, because Bambu Lab P1P's bed plate size (256 x 256 mm) is somewhat smaller, larger parts were still printed using CR-10 Smart's bed plate size (300 x 300 mm). For better print quality, calibration of the print settings was frequently performed. As a result, the construction of the scale model was completed using these two printers without outsourcing.

(a) Creality Cr-10 Smart (Source: www.creality.com)   (b) Bambu Lab P1P (Source: www.bambulab.com)

Figure 45:   **FDM 3D-Printers used in Construction of Scale Model**